D1535038

The Enigma of Time

The design on the cover is the Ouroboros. This is a serpent or dragon which in biting its own tail is symbolic of time, eternity and the continuity of life. The Greek inscription can be translated as 'The One, the All'. The counterbalancing of opposite tendencies as in the Chinese yin and yang is sometimes alluded to by having half its body dark and the other half light.

The Enigma of Time

compiled and introduced by

P T Landsberg

Faculty of Mathematical Studies,
University of Southampton

If you can look into the seeds of time,
And say which grain will grow and which will not,
Speak then to me, who neither beg nor fear
Your favours nor your hate.

Banquo in *Macbeth* Act 1, Scene 3

Adam Hilger Ltd, Bristol

British Library Cataloguing in Publication Data

Landsberg, P.T.
  The enigma of time.
  1. Time
  I. Title
  529′.01   QB209

ISBN 0-85274-545-1

Published by Adam Hilger Ltd, Techno House, Redcliffe Way,
Bristol BS1 6NX.
The Adam Hilger book-publishing imprint is owned by
The Institute of Physics.

Printed in Great Britain by J W Arrowsmith Ltd, Bristol.

Contents

List of Authors

A Aharony, Department of Physics and Astronomy, Tel-Aviv University, Israel.

P C W Davies, Professor of Theoretical Physics, University of Newcastle-upon-Tyne, UK.

P A M Dirac OM, FRS, Professor of Physics, Florida State University, Tallahasse, Florida, USA. Dirac was one of the founders of the science of quantum mechanics and was awarded the Nobel Prize for Physics in 1933.

Sir Ernst Gombrich FBA, Director of the Warburg Institute and Professor of the History of the Classical Tradition, University of London, UK.

Sir Fred Hoyle FRS, research Professor at Manchester University and University College, Cardiff; Visiting Associate in Physics, California Institute of Technology, USA. Hoyle was formerly the Director of the Institute of Theoretical Astronomy at Cambridge. He was awarded the Gold Medal of the Royal Astronomical Society in 1968 and the Royal Medal of the Royal Society in 1974.

P T Landsberg, Professor of Applied Mathematics, University of Southampton, UK.

A J Leggett FRS, Professor of Theoretical Physics, University of Sussex, Falmer, Sussex, UK. Winner of the Simon Memorial Prize of The Institute of Physics in 1981.

G N Lewis (1875–1946). Lewis was a well-known American physical chemist. He was co-author with M Randall of a famous text on thermodynamics whose first edition appeared in 1923.

P Morrison, Professor of Physics, Massachusetts Institute of Technology, USA. He is a member of the National Academy of Science and was awarded the Presidential award of the New York Academy of Science in 1981.

Y Ne'eman, Department of Physics and Astronomy, Tel-Aviv University, Israel.

W J Ong, Society of Jesus, Professor of Humanities, St Louis University, Missouri, USA, and a Fellow of the American Academy of Arts and Sciences.

R Penrose FRS, Rouse Ball Professor of Mathematics, Mathematical

Institute, Oxford. Penrose was awarded the Eddington Medal of the Royal Astronomical Society in 1975.

E Schrödinger (1887–1961). Schrödinger was one of the founders of the science of wave mechanics and was awarded the Nobel Prize for Physics in 1933. In 1927 he succeeded Max Planck as Professor of Theoretical Physics at the University of Berlin.

Preface

The main purpose of this volume is to demonstrate the physical scientist's struggle to understand the nature of time by the reprinting of relevant discussions from the period 1930–1980. In order to be considered for inclusion a scientific article had to be expository, written in English, reasonably self-contained, less than 30 pages long and somewhat less easily available than, say, an article from *New Scientist* or *Scientific American*. This ruled out many interesting philosophical approaches and popularisations, but this did not make the choice any the less difficult. In the end I went for thought-provoking articles; but I was influenced, too, by my desire to present the work of both outstanding scientists and stylists. Thus we have examples from G N Lewis and E Schrödinger who were masters of the art of exposition.

The illustrations, and the more general articles in the last part, are all evocative of the wider setting of our subject. Man has always been puzzled by time and has attempted to understand it—by poetry and art as much as by science.

In an effort to make this fascinating subject accessible to a wider audience I have supplied an Introduction, which gives an explanatory survey of many of the areas covered by the reprints, and a Glossary. Difficulties and obscurities remain. They are characteristic of a subject which is not fully understood, and the reader should therefore not be dismayed when he comes across them. The book will have fulfilled its purpose if it leads to an appreciation of the nature of the problems of time as faced by the physicist and of their wider philosophical implications.

I am indebted to colleagues at the University of Southampton, Tony Hey, Ray d'Inverno, Ron King and Ashok Pimpale, for their helpful comments on the manuscript and also to Dr D A Evans and Dr E Schöll (R.W.T.H., Aachen) at University College, Cardiff and Dr A Danielian, London.

Peter Landsberg

Acknowledgments

The Institute of Physics (Adam Hilger Ltd) and the compiler gratefully acknowledge the permission of the following to reproduce illustrations and copyright material.

The American Association for the Advancement of Science for *The Symmetry of Time in Physics* by G N Lewis (*Science* **71** pp569–77 (1930)).

The American Philosophical Society for the plate of a 100-gradation incense seal from *The Scent of Time* by Silvio Bedini (*Transactions of the American Philosophical Society* **53** part 5 (1963)).

The Australian National University Press for *The Asymmetry of Time* by Fred Hoyle (1965).

The Bodleian Library for the plate of a water clock from MS Bodley 270^b, fol.183^v. The following have assisted in connection with this picture: Dr D Vaughan (Science Museum, London), Mr C K Aked (Antiquarian Horological Society) and Professor S Bedini (Smithsonian Institution, Washington).

Cambridge University Press for *Singularities and Time-asymmetry* by R Penrose from *General Relativity—An Einstein Centenary Survey* ed S W Hawking and W Israel (1979).

Cambridge University Press and Mr E Bredsdorff for the illustration of a hodometer of the Han dynasty from *Heavenly Clockwork: The Great Astronomical Clocks of Medieval China* by J Needham, W Ling and D J Solla Price (1960).

The Courtauld Institute for *Time Smoking a Picture* and *The Bathos* by Hogarth and *The Triumph of Time* by Pieter Breughel the elder.

The Dublin Institute of Advanced Studies for *Irreversibility* by E Schrödinger (*Proceedings of the Royal Irish Academy* **A53** pp189–95 (1950)).

Israel Universities Press for *Time Reversal Asymmetry at the Fundamental Level—and its Reflection on the Problem of the Arrow of Time* by Y Ne'eman from *Modern Developments in Thermodynamics* ed B Gal-Or (1974).

The Museum of Modern Art, New York, for Salvador Dali's *The Persistence of Memory* (*Persistance de la Memoire*), 1931. (Oil on canvas, $9\frac{1}{2} \times 13''$. Collection, The Museum of Modern Art, given anonymously.)

The North-Holland Publishing Company for *Time Reversal Symmetry Violation and the H-theorem* by A Aharony (*Physics Letters* **37A** pp45–6 (1971)) and *Time's Arrow and External Perturbations* by P Morrison (from *Preludes in Theoretical Physics* ed A de Shalit *et al* pp347–51 (1966)).

Pergamon Press for *The 'Arrow of Time' and Quantum Mechanics* by A J Leggett from *The Encyclopedia of Ignorance* ed R Duncan and M Weston-Smith (1977, pp101–9).

The Royal Astronomical Society for *Black Hole Thermodynamics and Time Asymmetry* by P C W Davies (*Monthly Notices of the Royal Astronomical Society* **177** pp179–90 (1976)).

Springer-Verlag for *New Ideas of Space and Time* by P A M Dirac (*Die Naturwissenschaften* **60** pp529–31 (1973)) and *Thermodynamics, Cosmology and the Physical Constants* by P T Landsberg (from *A Study of Time III* ed J T Fraser *et al* pp115–38 (1973)).

The University of Illinois Press for *Evolution, Myth and Poetic Vision* by W J Ong from *Comparative Literature Studies* **3** pp1–20 (1966).

The University of Southampton for *A Matter of Time* by P T Landsberg (Inaugural Lecture, 1975).

The Trustees of the Wallace Collection for *A Dance to the Music of Time* by Nicolas Poussin.

The Warburg Institute for *Moment and Movement in Art* by E H Gombrich from *Journal of the Warburg and Courtauld Institutes* **27** pp293–306 (1964).

William Heinemann Ltd for *H G Wells Talking to his Younger Self* from *Observations* by Max Beerbohm (1924).

Introduction

1. Aims and Assumptions

It has always been felt that time, as treated in poetry or art, is a mysterious concept and that philosophical treatments fare no better. When one turns to the physicists, whose science is the most fundamental, one might expect some clarification and a consensus. One does indeed attain clarification, as this book shows, but one does not always get a consensus, as this book also shows. Those who open it hoping to find an integrated picture of the physicist's view of time will in fact be disappointed. If this book teaches us anything it is that in each branch of physics questions about time raise deep issues, for whose resolution physicists strive with patience and persistence—sometimes for decades and occasionally for centuries.

In this Introduction nine strands of the physicist's problems of time will be distinguished and it will be shown where and how the individual reprinted papers fit into the scheme. It is hoped that the reader will in this way obtain an appreciation of the problem areas involved. Although the approach is largely non-mathematical, it must be expected that with a topic of this nature conceptual problems will arise. Roughly speaking the later parts of the book are the more difficult ones. An exception is in Part D (the papers by Gombrich and Ong) which some readers will find easier to absorb, since we exchange the technicalities of physics for those of the arts. Although the papers of Parts A, B and C should be read broadly in sequence, Part D is independent. It can be read first by those who seek to be convinced of the ubiquity with which problems of time occupy the human psyche. In this last section the authors contrast the treatment of the moment and the exultations of the present with discussions of physical movement in art and the assimilation in poetry of historical change, giving many interesting examples. It is this contrast which has been a source of creative tension in the arts and in philosophy long before the emergence of modern science—but it has been taken over into science as well, as evidenced by the continuing interest in the paradoxes of Zeno which are noted in paper 13 and which still appear in the specialised physics literature of the 1980s.

Following these remarks about Part D the rest of this Introduction discusses the problem areas in physics which are dealt with in Parts A, B and C.

1

A formal definition of time already presents major obstacles, but it is adequate for the present purpose to regard it as defined by the clocks used to measure it. This is satisfactory since the earth itself is a clock which measures approximately a year in going around the sun, sundials go back at least to Roman times, and there are pendulum clocks and clocks of many other types which are mutually consistent.

Time passes in so unforgiving a manner that it is apparently uni-directional. We cannot really turn back the clocks, for some other clock will show how much time was needed to do so. In fact a main concern of this book is with this unidirectional nature of time.

Some assumptions underlie the whole discussion given here. One is that the time variable is rather like a straight line on which a point marked 'The Present' moves uniformly and inexorably. This picture yields in fact two characteristics of time which are normally (but by no means always) assumed, and which are granted in this book: that time is continuous and that it is one dimensional. We will not take up alternative views, but merely note here that it *could* be otherwise. First, as one divides an interval ever more finely by the use of better and better instruments one could discover a graininess such as scientists found in the cases of mass and energy. Time might pass in indivisible lumps. Scientists have learnt to distrust first impressions and, like Eddington, one can write a survey like this at one of two tables. The solid smooth table of common sense which undergoes a mysterious metamorphosis when it burns to ash. Or one can use one's scientific table—a seething conglomeration of vibrating atoms and mole-cules whose transformations upon burning are however scientifically tractable. It is conceivable that it could be likewise with time. The common sense time may be matched by a scientific and granular ('quantised') time.

Secondly, time might have several dimensions. In that case one would not be able to say, of two distinct instants of time measured at one place, that one is earlier and the other later. It could be that in one direction of time this relation is opposite to that in any other direction of time: A precedes B in one direction and follows it in another. These matters raise intriguing problems, which have been considered in the scientific literature.

A preliminary indication of the many different views of the physical explanation of the unidirectional nature of time can be obtained from the Appendix of paper 7. The existence of these divergent views is responsible for the fact that it is believed that there is no single scientific paper which may be said to 'solve' the problem of time. (If it existed, it would be rather specialised, and so would not be included here in any case.) *Scientific* papers which deal with the time problem in a non-mathematical manner are not abundant, but we offer here what is hoped will be a stimulating selection.

Several of the papers were published in relatively inaccessible places and are clearly worth putting before a wider audience.

2. The first strand: Time in classical physics

Consider first classical mechanics and electromagnetism, i.e. the mechanics of Newton and the description of electric and magnetic phenomena as known in 1900, and associated with the name of James Clerk Maxwell. The key idea here is that the classical laws do not distinguish a direction of time. In the language of the physicist: they are time-reversal invariant. This can be translated into the idea of applying an operator T (defined to reverse the direction of time) to any of these laws. This operation may result in finding the same law. If that happens the law is said to be time-reversal invariant. It is time-reversal invariant in a generalised sense if one finds a valid law, though not necessarily the original one.

This matter is introduced in the well-written paper by G N Lewis and it is also considered in paper 4. Lewis's article gives an excellent discussion of the problem of time as it stood in 1930, when the universe was still widely considered to be a static rather than an expanding system. (The second law of thermodynamics mentioned by Lewis is noted under *entropy* in the Glossary.)

We now add a few additional remarks by way of introduction. That we *know* a direction of time is clear; that the laws of particle physics cannot distinguish a direction of time is a most remarkable fact. It can be illustrated as follows. Suppose a hard billiard ball is reflected from the side of a hard billiard table and is then removed. We are faced by (1) the projection and removal of the ball, examples of external interference known as *boundary conditions*, and (2) the undisturbed motion of the ball according to the *laws of mechanics* in the interval between these boundary conditions. A series of photographs of this latter motion could not then be ordered in time by an inspector to whom they are shown. The ball might be travelling one way or in the opposite way. Like John Buridan's ass, which starved between two equal haystacks because he could not decide from which to eat, our inspector has no way of deciding the matter in the absence of the boundary conditions. The latter are imposed by man and hence can readily be changed. They are therefore not fundamental. The *laws* cannot be changed and so they *are* fundamental. But they are devoid of a time sense. Even more surprising, the same can be said of the laws as currently formulated in quantum mechanics for the encounter of electrons, protons, atoms, etc. They, too, know no time sense (see section 6 below).

Considering light or electromagnetic radiation next, one again finds time-symmetrical laws, even though the equations of mechanics (Newton) have now to be replaced by those of electromagnetism (Maxwell). The particles of light ('photons') are actually reflected from a mirror in just the way a billiard ball is reflected after collision with the side of the table.

In fact, there *is* an asymmetry on the billiard table: for example, balls come to rest because they do not roll smoothly without friction, are not

3

perfectly hard (or rigid), etc. These departures from ideality *do* bring an arrow of time into physical laws, but under the heading of thermodynamics (p. 7), and these effects cannot be invoked if one confines the discussion to particle mechanics. However, as it is an aim of physical theory to *reduce* all subjects to a form of mechanics, the reader may already suspect a difficulty: is this reduction not impossible if mechanics is time symmetrical? This is discussed in section 4.

Classical particle mechanics, equally remarkably, led to paradoxes of motion which have survived two millennia of argument and doubt. Perhaps the purest of these paradoxes (Zeno's paradox) refers to the arrow which, flying from A to B, is at rest at each instant of its travel and so is at rest at *all* times. One possible answer is that to cover a finite time interval one needs an infinity of these instants and the distance through which the arrow moves is therefore infinity times zero. This can have any value and hence the possibility of motion is re-established. This escape via mathematics does not satisfy those who say that there is nothing residing in an arrow which is at rest at an instant that could help it on to a neighbouring point. So one has to admit a definition of velocity at a point P as the ratio of distance to time interval in the limit when they both go to zero, and the centre of the arrow is at the point P. Then the arrow has both position and non-zero velocity at P. Scientists do not worry too much about these issues, and yet they contain valuable clues. Viewed positively Zeno is sometimes considered a precursor of the quantum theorists who now deny the possibility of knowing with unlimited precision the position and velocity of a particle.

3. The second strand: Time in subnuclear physics

Any plane figure if rotated by 360 degrees about its centre gives the same figure, i.e. it comes into coincidence with the original figure. However, one needs to rotate an equilateral triangle through only 120 degrees about its centre to obtain an identical triangle, and the rotation of a square by only 90 degrees about its centre gives an identical square. The reason why these operations are possible is that these figures have a certain *spatial* symmetry, and the operations described are therefore called symmetry operations. If a law goes over into itself as a result of T invariance we again speak of a symmetry operation: the law goes over into itself when the direction of time is reversed. Thus the operator T applied to a normally projected film results in the film being shown backwards. If the film refers to billiard-ball collisions, i.e. to classical systems, the reversed film does not violate the laws of classical mechanics and this illustrates the T symmetry of mechanics.

Other such 'symmetries' exist. For example, the 'parity' operator P changes the system into its mirror image, and the charge conjugation operator C changes the sign of the electric charge and certain other charge-like quantities (called strangeness and the like). Put differently, C changes a particle into its antiparticle (see *antimatter* in the Glossary).

In figure 1 we apply the operations P, C and T sequentially to an abstract chemical reaction when a new possible reaction is found: one passes from

$$A + B \rightarrow C + D \qquad \text{to} \qquad \overline{C} + \overline{D} \rightarrow \overline{A} + \overline{B}.$$

If this reaction does in fact take place, TCP symmetry is said to hold for the original reaction. One can apply this to subnuclear particles, such as protons (p), neutrons (n), electron–neutrinos (ν_e), electrons (e^-), etc and their antiparticles $\overline{p}$, $\overline{n}$, etc. For example, the neutron–proton conversion

$$n + \nu_e \rightarrow e^- + p$$

implies also the conversion

$$\overline{p} + e^+ \rightarrow \overline{n} + \overline{\nu}_e$$

of an antiproton. This requires the capture of a positron (e^+).

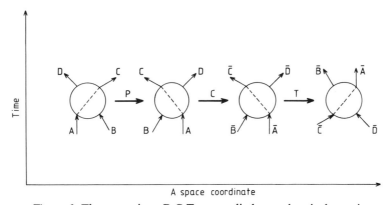

Figure 1. The operations P,C,T, as applied to a chemical reaction.

For many years P, C and T symmetries were assumed valid until an experiment proposed by Lee and Yang in 1956 to test for P symmetry, i.e. for parity invariance, in radioactive decays of nuclei, established its violation[1,2]. It caused consternation in the physics community. Now many other examples are known. For example, consider the radioactive decay of cobalt-60 nuclei at low temperatures. If a current is arranged to flow around the sample (figure 2(*a*)) so as to yield a downward pointing magnetic field, consider the electrons which are emitted in the opposite direction. In a mirror which is at right angles to the plane of the current loop, one sees both current and field reversed and the electrons are now emitted in the same direction as the field (figure 2(*b*)). Thus P invariance requires that the same number of electrons are emitted in both the direction of the field and in the

5

opposite direction. If the mirror is placed in the plane of the loop one finds the same result (figure 2(*c*)). Since the electrons are in fact emitted predominantly as in figure 2(*a*), the experiment establishes parity violation. This violation seems to occur only when *weak interactions* are involved (see Glossary).

The decay of certain neutral particles (kaons) furnished an even more exciting symmetry violation, that of CP, i.e. of P and C applied

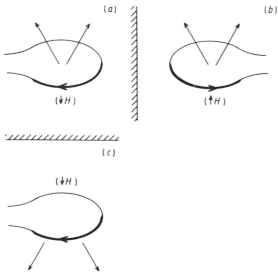

Figure 2. Mirror images (*b*) and (*c*) of the cobalt-60 decay experiment (*a*). The direction of the magnetic field (*H*), as derived from the direction of the current, is also shown.

sequentially[3]. Assuming that TCP, i.e. P, C and T applied sequentially, is a good symmetry, as seems to be the case, this implied a failure of T symmetry in a fundamental particle encounter. Here then, in the subnuclear field, one can find an objective indication of the *direction* of time, since the time inverse of the kaon decay process takes place at a different rate from the forward process[4]. These experiments (which again depend on the weak interactions) also enable one to give an absolute distinction between matter and antimatter.

This failure of T symmetry is exciting, but why it is so well hidden is an unsolved puzzle. It is believed to be unimportant outside the subnuclear domain. Its key significance lies in the fact that these decays, if incorporated in a cine film of (for example) billiard-ball collisions, would enable the observer to detect if the film is run forward or backward. They therefore provide an *arrow* for the time axis and provide a law for particle encounters which is not time-reversal invariant. Ne'eman's 1974 paper gives a succinct summary of the problems of the arrow of time and emphasises T, C and P symmetries. The usefulness of this paper resides in the fact that it deals briefly and clearly with the various possible arrows of time and the question of why they point in the same direction.

A system may be subject to laws having certain symmetries but the appropriate symmetric solution may not be exhibited by the system. This can happen in principle when the symmetric behaviour is highly unstable to disturbances. In this way the concept of hidden symmetries arises. It is not known whether P and CP symmetries are hidden in that sense.

Figure 1 shows that since TCP (rather than T) is a good symmetry for elementary interactions, one can regard TCP as a (weakened) time-reversal invariance which takes one from matter to the time-reversed processes in antimatter, or from antimatter to the time-reversed processes in matter. We return to this point at the end of section 4.

4. The third strand: Time in the reduction of thermodynamics to statistical mechanics

As Ne'eman summarises in his paper, there are many arrows of time. There is, for example, always the biological arrow: people add to their memories and grow old. Taking a *reductionist* view of biology, one needed to find an understanding of the arrow in terms of *physics* so that, as in a relay race, it could be handed on to biology. Thermodynamics was the only branch of physics having this understanding and it was stated in the entropy law (see Glossary) of thermodynamics, which can be regarded as the historically first arrow of time in physics (Clausius 1850, William Thomson, later Lord Kelvin, 1851). It led to the following problem. Suppose a gas of many molecules is in equilibrium and confined to one half of a partitioned box. On removal of the partition it can expand (or 'diffuse') into the larger volume and regain a new equilibrium state. Can it go back by itself? Physics says 'yes', for the molecular collisions which caused the expansion are time symmetric (the kaon decay of section 3 can hardly play a part). This is the extraordinary conclusion to be drawn from the previous section, and physics must give an explanation of the fact that the contraction of the gas into the half of the box it first occupied is never seen. Thus, although physics says it is possible, for a gas of many molecules it is so unlikely as to be practically impossible.

This return of an isolated system in equilibrium to any of its possible states after an appropriate lapse of time is called a Poincaré recurrence. One can see that this recurrence time for the above experiment can be quite short if the 'gas' has three molecules: there is a good chance that they will be in the original half of the box again and again. But this chance goes down as the number of particles is increased, and for a normal gas of 10^{18} molecules per cm^3 it is almost never seen. We have thus a recurrence of states and also fluctuations of the entropy in an isolated system. Schrödinger discusses this in some detail in his paper, written with his typical enthusiasm and verve, the usefulness of which is that it gives a very readable account of just these problems. They are also noted by G N Lewis and Paul Davies.

It is intuitively clear that if one has an infinite number of molecules in an infinitely large container one can make the Poincaré recurrence time infinitely long and one can banish the oscillating entropy. Calculations in this limit (the so-called thermodynamic limit) are therefore rather simpler, and the derivation of thermodynamics from statistical mechanics works, strictly speaking, only in this limit. How seriously must this limit be taken? There are two unresolved problems to be considered: Firstly whether or not it makes physical sense to consider an infinite system and, secondly, as pointed out in paper 7, the fact that an infinite system is liable to suffer *gravitational collapse* (see Glossary). It is then unstable and thermodynamic equilibrium is impossible.

Entropy, as all concepts of ordinary life like pressure, volume and weight, is a *macroscopic* concept. What is the corresponding *microscopic* quantity? Boltzmann found a suitable expression for it and it is called $-H$. To give the expression for it here is unnecessary (see paper 4). The fact that entropy has a tendency to increase for an isolated system is mirrored by the tendency of H to decrease. This tendency is called the *H-theorem* (see Glossary). It provides a link between macroscopic thermodynamics and the microscopic (i.e. the particle-collision) picture of statistical mechanics. It is always essential for a proper understanding to effect this further reduction of macroscopic physics to microscopic physics. The apparent incompatibility of one-way macroscopic ('irreversible') processes such as heat conduction and diffusion with the time-reversible molecular collisions is then found. It must then be explained why the contraction of a gas into a corner of its box (antidiffusion), or the generation of temperature gradients in a bar originally at uniform temperature (anticonduction) hardly ever occur!

This suggests that a distinction must be made between complex processes and simpler ones such as collision processes in classical or quantum mechanics. We therefore define 'weak' T-invariance of a complex process by the requirement that its time inverse (although perhaps improbable) does not violate the laws of the most elementary processes in terms of which it is understood. Thus, diffusion and heat conduction, while not T invariant, are at least weakly T-invariant. The difference between these two invariances could be said to lie in the statistics needed to describe a complex process in terms of simpler ones. On that basis what processes are *not* weakly T-invariant? The answer is: all those which have not as yet been decomposed into elementary processes, namely the complex processes of consciousness and of life generally. Hence one arrives at a definition of life *as a macroscopic process which violates weak T-invariance*. By introducing weak T-invariance we have chanced on a definition of life processes[5]! The definition in a sense codifies our ignorance. It is an unusual definition in that it focuses our attention on the possible temporary nature of our ignorance, since the definition will have to be revised if and when life processes and consciousness are explicable in terms of physics and chemistry. One should then be able, in principle, to create life from physico-chemical material and give a new 'constructive' definition of life to replace the above definition. These thoughts are briefly hinted at by Landsberg in paper 4.

We shall now attempt a partial summary of what has been found. For this purpose let time reversal in the TCP sense be applied to thermodynamics. As explained at the end of section 3, this yields antidiffusion and anti-conduction in antimatter. However, antidiffusion and anticonduction must be as unlikely for antimatter as they are for matter. Let us therefore rule them out. Since the operation of TCP on thermodynamic processes then yields inadmissible (or unphysical) processes, it follows that thermo-dynamics violates the TCP time-reversal invariance. Thus one may conclude that TCP time-reversal invariance applies to elementary processes, including kaon decay but it is violated in thermodynamics and statistical physics where only weak T-invariance survives, and even that is lacking in life processes.

5. The fourth strand: Time in statistical mechanics

Let us next send the gas back into its corner by reversing all molecular velocities at a certain instant. This can be done in a computer experiment, but not in real life. One then finds that the entropy decreases from these new and contrived sets of 'initial' conditions, as already envisaged by Josef Loschmidt (1876). This result is most unexpected according to the second law, and it has been called *'antikinetic'* for Loschmidt's paradox suggests behaviour opposite to that predicted by normal statistical mechanics (or *kinetic* theory as the study of molecular motions is called). It is here that the graph labelled $\beta = 0$ given in Aharony's paper is of importance, for it gives the results of precisely such a computer experiment (the details given there in equations (1) to (3) are needed only by the specialist). If small random errors are introduced into the calculation to represent, for example, the irreducible interaction with the outside world of a so-called isolated system, then this antikinetic behaviour is suppressed. A similar suppression is produced by using an interaction which is not invariant under time reversal ($\beta \neq 0$), an example being the interaction in kaon decay. Both changes encourage an increase of the entropy even after Loschmidt's velocity inversion, showing how delicate and unstable the initial conditions are which result from this inversion.

9

Philip Morrison deals in more detail with just this question of an external disturbance in paper 3, and that is why it is included here. He notes that only the universe itself can escape these remaining interactions, and that gravitational interactions (against which there is no shield) are important in this connection. Fred Hoyle (paper 5) also argues that the arrow of time comes from outside the system and explicitly rejects thermodynamics as providing an arrow of time. In this latter respect Hoyle is in a minority, but his reasoning in the first section of part II of his paper is of interest. Roger Penrose, by way of contrast, points out that in connection with the arrow of time 'entropy is crucial'[6], a view shared by the present writer.

If one belongs to a school of thought which does not wish to import the arrow of time from outside the system other possibilities are available, of which we shall note three.

(i) One could add to the laws of dynamics a statistical law from which the direction of irreversible processes can be derived[7]. Care must be exercised though, for the gas will certainly be seen to go back occasionally if the number of particles is small.

(ii) It can be noted that because one's apparatus is always imperfect, a large system can usually be located in only one group of many somewhat similar microstates or quantum states of *phase space* (see Glossary). Such an imperfectly focused state is called a macrostate. The process of passing from microstates to macrostates is called *coarse-graining* (see Glossary). The second possible addition to the laws of physics goes as follows: When a group of macrostates corresponding to equilibrium has an overwhelmingly larger probabilistic weight than any other such group, the system will be found in this group with a corresponding overwhelming probability, and in others with corresponding low probability. One can show that the entropy, when expressed in terms of macrostates, would increase in the above diffusion-type processes. This is essentially principle **P** which is discussed in paper 4. As stated in the Glossary, the fine-grained entropy of a system is constant in time, so that coarse-graining is in some sense responsible for entropy increase.

(iii) One may take the sudden velocity reversal seriously, assume it to be experimentally possible, and note that the system is not isolated at the instant of reversal. Its entropy could therefore drop at that instant (although this has not been generally agreed), to increase once more when the system is isolated again[8,9]. The Loschmidt reversal *does* then lead to a lowering of the entropy but it also remains true that entropy is non-decreasing for an isolated system.

10

I. *A case of time reversal.* As a sign of the Lord's wish to help him in his illness Isaiah had offered to King Hezekiah a 'practical' demonstration—namely the movement of the sun's shadow forward or backward by ten degrees. Not unnaturally King Hezekiah felt that this demonstration would be more convincing if the shadow actually moved backwards. In this way the backward movement of time became one of the biblical miracles [Kings 2:20, Isaiah 38:3]. The appearance of this backward movement of the sun can be simulated by pouring water into an appropriately designed bowl sundial and using the refractive properties of water. It is not possible to simulate it by a water clock—which was in any case introduced after biblical times.

The miniature shown here is taken from a Latin manuscript in the Bodleian Library, and dates from about AD1285. The manuscript describes the miracle and illustrates it by depicting a water clock. It is not explained how the water clock actually works (a factor which is still a matter for discussion). The juxtaposition of King Hezekiah's story with a water clock is, of course, an anachronism. This clock was presumably a monastic clock made to sound at certain prearranged times to help the sacrist. The hodometer shown in figure VII is a precursor of this type of 'interval timer'. For more details see the article by C B Grover in *Antiquarian Horology* **12** 160 (1980).

6. The fifth strand: Time in quantum mechanics

Instruments such as telescopes, balances, etc, are within a few factors of ten, of human size. Thermodynamics and statistical mechanics can thus be applied to them, even though the systems studied fall into three broad classes: (*a*) those of the same size (classical physics), (*b*) those of much greater size (astrophysics) and (*c*) those of much smaller size (quantum physics). In all cases an instrument is activated on being connected to and disconnected from the system. The activation energy subtracted from the system is dissipated by the measuring instrument as it returns to its original state, ready for the next measurement. Even a null experiment requires neighbouring non-null observations to check that the apparatus actually works. Thus the measuring process is associated with entropy increase. The activation and deactivation of the instrument, even if fully automated, are operations the time inverse of which are *hardly likely*. Such time-inverted operations would result in an emission of what was originally the incoming signal. It is true that with amplifiers, induction coils, etc, in the measuring circuit, several irreversibilities are heaped upon each other, but a measurement is still only an irreversible process of the *weakly T-invariant* variety. This is how the above term *hardly likely* must be interpreted. We now turn specifically to class (*c*).

Time enters quantum physics *in the first place* through its time-reversal invariant equations (such as Schrödinger's equation). Now in quantum measurements the connections and disconnections of the apparatus mean, for example, that an electron initially represented by a spread-out wave may eventually produce a highly localised oscillation on a screen. Its wavefunction is said to have 'collapsed'. In some manner not describable by standard quantum mechanics, all alternative potentialities for the states of the electron have become inaccessible. Many attempts exist to give a satisfactory description of these effects[10]. They include the view that the reduction of the wave is an irreversible thermodynamic process, the view that the measurement problem is insoluble within quantum mechanics, a recent reinterpretation of the entropy law that is believed to incorporate the collapse of the wavefunction within quantum theory[11], etc. Thus questions of time enter quantum physics here in a *second* and, as we shall see, most problematical manner: (1) Does the wavefunction apply to individual systems or to ensembles of systems? (2) If the latter, is quantum mechanics a form of statistical mechanics corresponding to a subnuclear mechanics, yet to be discovered? If so, can this new mechanics be constructed by adding new or *hidden variables* (see Glossary) to those conventionally used? (3) Quantum theory, while not allowing systematic signalling at speeds faster than light, implies that causal influences *can* spread at such speeds. Thus one finds dependences of parts of systems on each other even if these parts are so widely separated that causal influences between the parts, if conceived as

13

propagating, would have to travel faster than light. How is this so-called *non-locality* to be interpreted?

In the *Copenhagen interpretation*[12] (see Glossary) one is satisfied that quantum mechanics applies to individual systems so that question (2) does not arise. The statements in (3) have experimental support[13,14] and raise deep questions about time especially as regards the speeds and the methods of propagation of physical influences. The famous discussion between Bohr (1885–1962) and Einstein (1879–1955) about the nature of quantum theory was essentially about these matters, but it did not have the benefit of the relevant experimental results (which have come in only since 1972) or of the theoretical clarifications based on Bell's inequality (1964). The position roughly identified with Einstein who wanted physical theory to be *local* ('Einstein locality') has thus been rather undermined by recent work. This seems to favour Bohr's position which was broadly that the description of physical reality by quantum theory is complete. Nonetheless the problems raised in (3) are most perplexing and have recently, after a slow beginning, been discussed much more widely in the physics community[12-19]. Action-at-a-distance or even the consciousness of the observer *may* have to be utilised in order to give a consistent and intuitively clear interpretation of all aspects of quantum mechanics, including the collapse of the wavefunction.

Leggett's paper gives a helpful introduction to this group of questions. He first touches on matters noted here in section 4, emphasising the ability of human beings to prepare ordered states (as initial or boundary conditions). These states become disordered *later*. One could of course argue, however, that the direction of time in statistical physics is not related to boundary conditions imposed by scientists in their laboratories. It is impersonal. The moving shadow of a boulder on the surface of a planet will initiate heat conduction in the normal sense whether or not it is observed by an intelligence. It can be regarded as being due to the statistical weight of the macrostates of the system, by Landsberg's principle **P** (paper 4). The reverse processes involve macrostates which, although accessible, have a small probability. A hot spot *could* be produced by neighbouring atoms sending kinetic energy in that direction in conspiratorial unison, but this is highly unlikely—as unlikely as the bath water molecules conspiring to send a drop back into the tap. Thus the statistical derivation of an arrow of time need not be *just* the result of a survey of the technological competence of experimenters for arranging delicate and unstable initial conditions[20]. Nonetheless the human mind *is* involved, as Leggett insists several times. The most obvious involvement is the one which is most neglected, namely that even the processes which occur *without* human participation are analysed and discussed by us, and so have been filtered by the human brain. Products of the human brain enter a second time through the instruments assumed to be available: the better the instruments, the higher the entropy found, as more states can then be distinguished.

In connection with Leggett's emphasis on consciousness, we now pass to life and intelligence. Consider a uniform N-molecule gas in a container. A partition with an orifice in it is situated in the middle of the gas, and an intelligent being lets molecules through it in one direction only, so as to render one compartment eventually empty. It does so, apparently without doing work, by just opening or closing the orifice, depending on which way the approaching molecule is travelling. This is again antidiffusion, and is opposite to the diffusion process envisaged at the beginning of section 4. There is therefore apparently a drop in entropy, in contradiction to the entropy law. This puzzle of the *pressure demon* can be resolved by noting that the demon needs a torch of some sort to see the molecules, and this can be shown to increase the entropy of the systems involved to more than it was in the first place.

The above demon is related to a demon introduced by Maxwell who, being accustomed to classical systems, assumed implicitly that the energy required for any observation can be made negligibly small. However, for the longest wavelength λ that can just resolve a molecule, we know now that at least one photon of energy hc/λ is needed, and so the energy *cannot* be made negligibly small. Maxwell's demon produced by analogous operations a temperature difference in a system which was initially in equilibrium. Maxwell's demon died at the age of 62 (when a paper by Leo Szilard appeared)[21,22], but it continues to haunt the castles of physics as a restless and lovable poltergeist.

7. The sixth strand: Time in special relativity

The best known of the odd properties of time is the so-called clock paradox, which is not a paradox as Hoyle clearly explains in paper 5. Before reading his paper the entry under *light cones* in the Glossary should be consulted.

Special relativity also leads to the notion of a *horizon* (see Glossary) which is important in general relativity. Suppose space traveller A is accelerated away from a man Z who is not accelerated. Then A's velocity relative to Z increases, he always feels a force acting on his body, and he will approach the velocity of light ever more closely. Z can contact him at first, but there will come a time when the fastest signal can no longer reach him: A has then crossed what may be called Z's radar horizon. Conversely, A is aware only of events which lie within a surface of space—time called his event horizon. It can thus be seen that in order to understand this concept it is essential to accept that no material object can be accelerated beyond the velocity of light c.

The fastest information transmission is thus via light rays. What shocks has physical time in store for us should particles exist which travel faster? Such theoretical particles are called tachyons or superluminal particles. As

15

particles cannot be accelerated or decelerated through the light velocity barrier, tachyons must be *intrinsically* superluminal. One shock is that, *if* tachyons exist, and *if* they can be used for signalling, then one has secured too high a speed of information transmission! It is so fast that things can be arranged in such a way that one can signal to one's own past. A £1000 bet lost by our man Z then becomes an event of which prior notice can be given to Z by the use of tachyons, so that he does not place his bet. Of course, this seems contradictory and most people feel that this kind of 'anti-telephone' must be ruled out. The defenders of the (undiscovered) tachyons maintain that for various reasons these particles could *not* be used for signalling, thus banishing the causal anomaly[23]. The second shock is that tachyons would have to have an imaginary rest mass. Further, their total mass energy *decreases* as the velocity goes up relative to our unaccelerated man Z. Since experiments yield real numbers, it is vital that the rest mass should not be observable even in principle. This is indeed so since, according to standard theory, there exists no *inertial frame* (see Glossary) in which the tachyon is at rest[5]. This is quite different from a normal particle of non-zero rest mass. There always exists an inertial frame in which it is at rest, its mass is then positive, and as it speeds up its mass *increases* thus making it increasingly difficult to accelerate it further. Certain astrophysical objects do *appear* to expand with a velocity in excess of the velocity of light[24], but this is a problem for which explanations within standard relativity theory are being sought.

There are two further interesting points to which allusion may be made here as they are relevant to this area of work.

(i) In 1965 a discovery was made which leads us to the belief that the universe is bathed in radiation which originated after the big bang. It is almost identical to the radiation from a black body at about 2.7 K, and has been rather difficult to explain in terms of the steady-state theory of the universe, which is therefore currently out of favour. It must be remembered that Hoyle's spirited advocacy of the steady-state theory in part II of his paper was written before this discovery was made. It is the only part of this volume which deals with this model. As it may come back one day in a modified form, it was felt desirable not to neglect it in the present work.

(ii) It appears that our galaxy, seen in the sky as the Milky Way, has the small velocity of about 600 km s^{-1} relative to the black-body background radiation[25]. The preferred frame of reference used in this measurement is that in which this background radiation appears isotropic. (For our galaxy it is *almost* isotropic.) There is no contradiction here with the teaching of special relativity that the general laws of physics do not depend on the inertial frame in which they are formulated, since for any given system there often does exist a frame of reference which is more relevant than others. For a thermodynamic system, for example, it is the frame in which the box which encloses it is at rest.

16

II. *A causal anomaly*. H G Wells Talking to His Younger Self, by Max Beerbohm. This is an allusion to Wells' very popular *The Time Machine: An Invention* (1895, London: Heinemann).

8. The seventh strand: Time in general relativity theory and gravitation

We now come to the most recent developments which are discussed here by Roger Penrose and Paul Davies. They raise quite difficult matters which require the introductory remarks given below.

Each atom has, as a kind of finger print, a characteristic series of spectral lines it emits, just as a set of organ pipes has *its* typical notes. One measures them in both cases in terms of frequencies. A low frequency, or bass in acoustics, corresponds to a low frequency in the red or infra-red part of the spectrum in optics, and both these periodic disturbances have a comparatively long repetition time. A high frequency, or treble, corresponds to a spectral line in the blue or ultraviolet and both have a comparatively short repetition time. Repetition times remind one of the time between ticks of a clock—another periodic phenomenon. In section 7 A's clock has its repetition time lengthened relative to an identical clock used by Z and hence less time passes for A than for Z. A's clock appears to be slowed down (this is the phenomenon of time dilatation). The clock on a receding source also appears to be slowed down relative to an identical clock at rest, and the light emitted by the source is therefore shifted towards the red end of the spectrum. This red shift is well known as the Doppler effect (p. 65).

Turning to analogous effects in the general theory of relativity, note that it attributes gravitation to a deformation of the geometry of space–time which endows it with curvature. One finds again that time spans depend on the frame of reference. Thus the collapse of a star to a black hole state takes a finite time when one moves with the stellar surface. For an outside inertial observer, however, the flow of time is arrested at the event horizon (see Glossary) and the collapse takes an infinite time. Similarly, one finds a red shift (see Glossary) due to emission of the light in a gravitational field—just as a ball thrown up into the air loses kinetic energy, so a photon loses energy, and hence suffers a red shift, as it climbs against gravity. Similarly, there is a gravitational slowing of clocks. All these effects are confirmed by the study of spectral lines emitted by the sun, white dwarf stars, and other strong gravitational sources, and they are discussed in books on general relativity theory.

The black hole concept will be introduced next. As one approaches a source of a gravitational field the velocity with which one has to project a stone away from it so that it will not fall back (the escape velocity) increases, and in some cases it can reach the velocity of light before one has reached the source itself. Such a source is a black hole and the surrounding surface with escape velocity c is its event horizon. Since objects fall into it and nothing is emitted from the inside, these processes furnish another arrow of time. As any object radiates—in the simplest cases as the fourth power of its absolute temperature T—this suggests that the 'classical' black hole just described is

19

at the absolute zero with the event horizon representing a *sink of information*. A black hole has typically a mass M, a charge Q and an angular momentum J, and no other parameters which might represent the nature of the material it has swallowed (this is the theorem that 'a black hole has no hair'). It is like a thermodynamic system without *microscopic variables,* but with macro-variables M, Q, J. If this were all one could say, the black hole

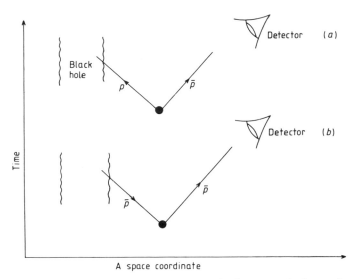

Figure 3. A particle p travelling forward in time as equivalent to its antiparticle $\bar{p}$ travelling backward in time.

processes noted would satisfy the definition of life given on page 9: neither black hole nor life processes would be understandable in terms of time-reversal invariant elementary processes. However, the variables of quantum field theory have still to be introduced, and they supply just these elementary processes.

The classical picture is then amended, and vacuum is seen to be a source of great activity: particle–antiparticle pairs form, separate, come together, and decay. Creation and annihilation occur at many space–time points, but the pairs $(p, \bar{p})$ cannot be observed (they are therefore called 'virtual'). Energy conservation implies that one partner can have negative energy and because of the nature of the black hole potential it can fall into the black hole; the other particle, now unable to annihilate, is then observable (figure 3(a)). The negative energy particle may be conceived as its antiparticle emitted by the black hole and travelling backwards in time to be scattered as a particle into the positive time direction at the position of materialisation (figure 3(b)). To pay for this the energy and mass of the black hole decrease (Hawking effect),

20

and the enlarged system which includes a region outside the event horizon therefore radiates. In fact, it behaves rather like a standard (black) radiator at an absolute temperature[†]

$$T = \frac{hc^3}{16\pi^2 GMk} \text{ (quantum mechanics)} \xrightarrow{h\to 0} 0 \text{ (classical theory)} \quad (1)$$

which does go to zero in the classical limit $h \to 0$. The mass and the event horizon of the black hole shrink as a result of this evaporation-like process, which incidentally, can be described mathematically in an approximate simple way[26,27]. Indeed, it is clear that the temperature T rises as the evaporation proceeds (since M becomes smaller), and this in turn increases the evaporation rate which is proportional to T^4. Black holes have large entropies (A being the area of the event horizon)

$$S = \frac{\pi c^3 kA}{2hG} = \frac{8\pi^2 GkM^2}{hc} \text{ (quantum mechanics)} \xrightarrow{h\to 0} \infty \text{ (classical theory) (2)}$$

and the merging of two black holes of masses M_1 and M_2 implies an entropy increase since $(M_1 + M_2)^2 > M_1^2 + M_2^2$. These formulae assume that the black hole is of the Schwarzschild variety, i.e. $Q = J = 0$, and show that general relativity leads to thermodynamic properties in a most unexpected and suggestive manner[28]. But the thermodynamics is not quite of the usual type in which the doubling of the system leads to a doubling of its entropy. This departure has rather striking consequences, of which the negative heat capacity of black holes is one[29]. This means that as a black hole loses energy (i.e. mass) its temperature actually rises, as may be verified from equation (1).

Penrose's and Davies' papers can both be read against the background sketched above. Both require use of the Glossary for such terms as *space–time diagram, Penrose diagram, Riemann tensor, singularity,* etc. These papers are important in that they familiarise us with current speculations. They are at the boundaries of knowledge and will be understood in detail only by specialists. Penrose speculates that the time inverse of black holes (white holes) do not exist, this is also discussed by Davies; they ask whether the big bang of cosmology was of low entropy and in what sense. Penrose in particular considers how to associate the entropy in the gravitational field with the tensors of relativity. Both papers deal with questions of both general relativity and of cosmology.

Standing back from the details of these two papers, our earlier discussion leads us to three problems, which will be raised here, but which cannot be resolved as yet.

† In equations (1) and (2) the following symbols are used: c, velocity of light in vacuo; h, Planck's constant; G, Newton's gravitational constant; k, Boltzmann's constant; π, ratio of circumference to diameter of a circle (approximately 3.14).

(i) The black hole entropy formula given above is rather striking. Where is the coarse-graining which was considered so important? One may conjecture that when there is no possibility of the entropy remaining constant on the merging of two identical systems (as was shown above for the black hole case) then there is no possibility of ideal reversibility, and so coarse-graining is not needed. The theorem of a time-independent fine-grained entropy would not, after all, appear to be applicable in these cases.

(ii) In this discussion we have lost time symmetry which, kaon decay apart, is involved in the basic theories! There are three ways of saving it. (*a*) The reverse process to black hole formation and evaporation must occur, i.e. disordered photon and particle radiation implode by an anti-Hawking effect and evaporate by the spewing out of ordered matter. The latter phenomenon may be regarded as a white hole division with decrease of entropy. This is no worse than antidiffusion (section 4) and must be allowed. However, the weight of probabilities must be overwhelmingly against it by virtue of some generalised principle **P**. Black holes, and the absence of white holes, are no more remarkable as regards time asymmetry than thermodynamics, and both are weakly T-invariant. (*b*) One identifies the Hawking evaporation as the time reverse of black hole formation, by arguing that once every blue moon the black hole evaporation can produce the sort of organised matter (books, tables, piano) which perhaps fell into the black hole. The white hole concept is then not needed. (*c*) Black holes do not exist, in which case the whole problem disappears. At present all three cases are possible, though (*c*) has little support.

(iii) Gravitational entropy can be approached via the statistical mechanical (or information theory) expressions for $-H$, utilised in Aharony's computer experiments discussed on p. 143. (This procedure can also be useful in analytical work which does not require computers[11.30].) Equilibrium thermodynamics is not much use, however, since gravitational systems can attain equilibrium only by virtue of non-gravitational forces, such as molecular binding forces in stars. The identification of a black hole entropy formula (2) as a thermodynamic (rather than a statistical mechanical) concept was therefore a great success. After all, purely gravitating systems show no equilibrium states and this precludes the use of a common thermodynamic trick—namely calculating a change of entropy between two equilibrium states A and B by choosing *any* convenient path that goes only through equilibrium states. The question now arises as to whether or not there is a more general approach to the calculation of *thermodynamic* entropies of gravitating systems of which the black hole calculation leading to (2) is but a special case. No universally accepted answer to this question has yet emerged.

9. The eighth strand: Time and the large number hypothesis

Whenever one wishes to improve a current theory it is difficult to see in what respect it is too general and in what respect it is too restricted. This holds true of general relativity. One disturbing fact is that many solutions of the Einstein equations of general relativity possess causal anomalies of the type discussed on p. 16 (such solutions admit so-called 'closed time-like lines'). These anomalies are of special interest because here it is physical objects themselves, not just signals, which would be able to travel into the past. One feels that they should not be allowed *by the theory,* rather than being dismissed out of hand, as it were, i.e. by a later edict. In this sense the theory is too general. On the other hand, the theory employs Newton's gravitational constant G as a time-independent quantity whereas it has been suggested that it may change very slowly over a time scale of billions of years. In that sense the theory may be too restrictive.

The large number hypothesis, and the reason for it, is spelt out in simple detail by Landsberg in paper 7, which is one reason for including it in this collection. It is also here that Dirac's paper is important. He explains in his famous simple style that he proposes to use two *metrics* (see Glossary), one for the cosmos the other for atomic systems, instead of relativity's single metric for both. To discuss this paper, recall the *steady-state model* of the expanding universe (see Glossary). This kept the density of matter in a steady state by assuming continuous creation of matter. As was seen on p. 16 astrophysical evidence no longer favours this model. Dirac's two-metric approach to the expanding universe originally also invoked creation of matter, as is clear from his paper. In its newer version, however, mass is conserved[31].

Could other 'constants' depend on time? The answer so far is that such dependence is zero, or is so small as to be beyond measurement[32].

10. The ninth strand: Time in cosmology

A cluster of profound questions as to the nature of time arises in the study of cosmology. When one imagines solar systems and galaxies smeared out into one fluid of matter, or two fluids of matter and radiation, one has a model which fulfils the cosmological principle (i.e. it is *homogeneous* and *isotropic,* see Glossary), and an appropriate expanding version seems in broad agreement with what is observed. The theoretical description of this state of affairs is provided by the so-called Friedmann models. These can exhibit expansion, oscillation or contraction; the first two alternatives alone are believed to be options for our present universe. This (initial) expansion is due to a *boundary condition* which is a big bang. The *laws* which operate are those of general relativity and electromagnetism and they are T invariant.

Life has been eliminated by the smearing out process so that these model universes are certainly weakly T-invariant.

Paper 4, *A Matter of Time,* has been included because it gives in simple language a brief statement of several current ideas in cosmology and because it compares these with older ideas. It shows in particular why there is a stark choice between an ever-expanding and an oscillating universe. Using present theories, the numerical values of the parameters which determine whether our universe is ever-expanding or oscillating are known to lie within certain ranges, but this knowledge is not precise enough to enable us to make a clear decision. Cosmologists are of such a turn of mind that they may be expected to ask 'what is the feature of the universe which makes it so hard for us to know which option applies?' We are on highly speculative ground, but one answer to this question is given in paper 7.

We now make four additional points which link up with the papers reprinted in this volume, or which illuminate them. Two deal with oscillating model universes and two with the large-scale properties of the models.

(i) Oscillating universes *which are time symmetrical* about the state of maximum expansion have been discussed[33-36]. But they require there to be a strong statistical connection (a *correlation*) between photons emitted in different directions by the big bang. The same would have to apply to other particles coming out of the big bang, and these connections would have to be preserved into the contracting phase as discussed by Penrose and Davies. Small disturbances could rapidly destroy the decrease in entropy in the contracting phase just as the rounding errors did in the computer experiment described earlier. To put it differently, very specialised boundary conditions would be needed for these universes. It is more reasonable, therefore, *not* to assume a time symmetrical universe. One would then expect randomness in the big bang, and one could attribute to it observed facts such as the absence of advanced electromagnetic radiation (which is discussed here by Lewis, Hoyle and Penrose) and the isotropy of the background radiation, which was noted on p. 16. For other applications of this assumption we return to statistical mechanics. Here it can be successfully assumed that *for an equilibrium system* subject to given constraints all the states which are possible are equally likely. This is just the randomness assumption needed here. And its justification? Well, one would have to regard it as a statistically reasonable procedure when one wants to take an unbiased stance in the absence of a great deal of information. (Once one is *away from equilibrium* the transition rates between pairs of states are no longer equal and some states are therefore preferred.)

(ii) One can discuss the change ΔE in energy and ΔS in entropy of an *oscillating* model universe of matter plus radiation as a result of one cycle. From an initial high thermal equilibrium temperature of matter and radiation, the effect of expansion in the absence of any matter—radiation

24

interaction is that the matter temperature falls more rapidly than that of radiation. As there is no interaction everything is symmetrical about the state of maximum expansion and $\Delta E = \Delta S = 0$. Interaction will slow down the fall in matter temperature and hasten the fall in radiation temperature. The pressures of these two components are affected by this, and it is found that the cosmological models behave remarkably like thermodynamic systems worked irreversibly and cyclically by external forces, as emphasised by Davies in his paper. In this way the oscillating model can be seen to be weakly T-invariant because the systems of statistical mechanics have just this property.

If there is *indefinite expansion* the changes in the universe are more permanent: galaxies and solar systems condense out of the initial hot soup in a universal process of gravitational clumping, and remain condensed out. The evaluation of the entropy contributions from this process is still a matter of current research[37] and it is discussed by Penrose and Davies.

(iii) The universe appears to be very uniform and isotropic and cosmologists ask why this is so. In fact various mechanisms are known which may have damped out early irregularities. Also, it is strange that different observers, at the same time after the big bang, can see the same type of smeared-out universe, even though these regions may not have been able to influence each other (if the observers are outside each other's light cone). Why are they so similar in spite of this lack of contact? Some randomness and isotropy in the big bang or soon thereafter is again clearly indicated[38], and attempts have been made to explain it.

The same idea may be applied to other singularities such as black holes. Thus the black-body radiation, which they emit at some absolute temperature, is just the 'most random' one for the given energy. One appears to have the general principle that a singularity emits with equal probability each configuration of particles compatible with its external constraints[39]. This leads us directly to what Davies in his paper calls the Randomicity Principle, sometimes called the 'principle of ignorance'. Thus statistical thermodynamics comes into its own, just where one might have expected it, in the case of singularities, where a full microscopic description is inaccessible. It is intriguing to note also that at black hole singularities, where space—time descriptions break down, the thermodynamic equilibrium description of black holes (section 8) which involves neither space nor time, is still useful.

(iv) This appearance of 'order' in the form of galaxies etc, in the universe is contrary *at first sight* to what the entropy law would suggest. But one must not forget that gravitational effects are always keeping the universe away from equilibrium, so that the entropy law is irrelevant in as far as it governs the linking of *equilibrium* states. We know that ordinary finite systems, which are driven away from equilibrium, display a reservoir of structure

25

which is not expected when one looks merely at their equilibrium states[40,41]. It would thus be rash and erroneous to say that as the universe is a developing structure, its entropy must be decreasing. Quite apart from this consideration, we know in addition that if the number of observable states of a system goes up, there is the possibility of entropy increasing while the order, defined in a suitable technical manner, *is also increasing*. This is the order-from-growth concept[26]. It can also be applied to biological evolution where the evolution of species for example appears, *at first sight,* also to defy the second law.

Singularities have been very important in the above account, but there is no certain evidence of their existence. We therefore conclude by giving a simple mathematical argument suggesting why one would expect them. Consider a volume of space determined by a surface drawn through specified galaxies in an expanding universe. One can easily verify[42] that the linear dimensions grow with time t. This can be modelled in Newtonian physics by a scaling factor $R(t)$, which scales up the distance between any two particles. It satisfies the equation for the energy E inside the surface

$$E = \text{kinetic energy} + \text{potential energy} = a\dot{R}^2 - b/R$$

where a and b are constant and $\dot{R}$ is the rate of increase of R. By differentiation one finds, since b/a can be shown to be proportional to the density ϱ of the energy,

$$\dot{R} = -\frac{b/a}{2R^2} \propto -\frac{\varrho}{R^2}.$$

This shows that the scale factor is concave towards the time axis (figure 4). Hence, if the theory applies at all times which have positive $R(t)$, there exists at least one time t_s for which $R(t_s) = 0$. It follows that density and temperature diverge at that time. The mere requirement that ϱ be positive is thus sufficient for the occurrence of a singularity at t_s. This result goes over into a related, but somewhat more complicated, theorem in general relativity[43,44].

Concluding remarks

Fifty years have elapsed since Jeans wrote the following[45].

'These considerations . . . have led many physicists to suppose that there is no determinism in events in which atoms and electrons are involved singly, and that the apparent determinism in large scale events is only of a statistical nature. . . . To-day science has no longer any unanswerable arguments to bring against our innate conviction of free-will. On the other hand, she gives no hint as to what absence of determinism or causation may mean. If we, and nature in general, do not respond in a unique way to external stimuli, what determines the course of events? If anything at all, we are thrown back on determinism and causation; if nothing at all, how can anything ever occur?

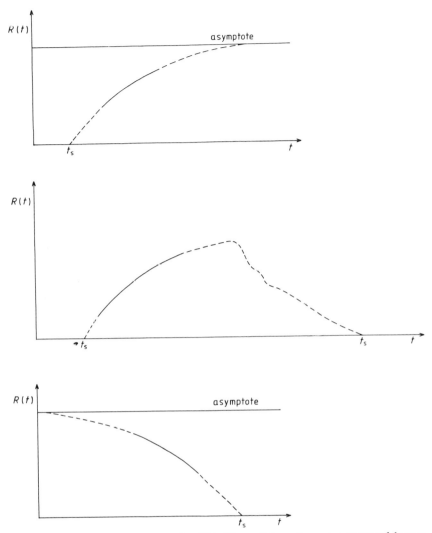

Figure 4. Some concave curves for $R(t)$. The solid portions are assumed known and the dashed parts are inferred possibilities from concavity. If the scale factor $R(t)$ is concave as here then it vanishes at least at one instant of time.

As I see it, we are unlikely to reach any definite conclusions on these questions until we have a better understanding of the true nature of time. The fundamental laws of nature, in so far as we are at present acquainted with them, give no reason why time should flow steadily on: they are equally prepared to consider the possibility of time standing still or flowing backwards. The steady onward flow of time, which is the essence of a cause–effect relation, is something which we superpose on to the ascertained laws of nature out of our own experience; whether or not it is inherent in

27

the nature of time, we simply do not know . . . the theory of relativity goes at any rate some distance towards stigmatising this steady onward flow of time and the cause—effect relation as illusions; it regards time merely as a fourth dimension to be added to the three dimensions of space.'

These views, hinted at in section 7, should be compared with those of the period 1930–1980, presented below, which show that considerable progress has been made. Future advances in this area will depend not only on further small forward steps but perhaps also on a major reconstruction of ideas to remove a problem of self-reference. Such problems, illustrated by the statement 'I am a liar', are difficult to handle. The problem is this—all our understanding depends on the nature of our brain. The brain in the form of 'the observer' enters a second time, for example in connection with quantum mechanical measurement. We are stumbling here because we do not know how to introduce the observer properly. It looks as if we need a better understanding of the brain which itself furnishes the machinery of all understanding. We seem to be in the position of a snake biting its own tail (a symbol of eternity in figure VI). If we make progress here we should be able to assess the view that the passage of time may be a figment of the human imagination—devised subconsciously to make up for another characteristic of the human brain, namely, that it is still a very weak instrument.

References

1 Lee T D and Yang C N 1956 Question of parity conservation in weak interactions *Phys. Rev.* **104** 254

2 Wu C S, Ambler E, Hayward R W, Hoppes D D and Hudson R P 1957 Experimental test of parity conservation in beta decay *Phys. Rev.* **105** 1413

3 Christenson J H, Cronin J W, Fitch V L and Turley R 1964 Evidence for the 2π decay of the K_2 *Phys. Rev. Lett.* **13** 138

4 Kabir P K 1970 What is not invariant under time reversal? *Phys. Rev.* **D2** 540

5 Landsberg P T 1972 Time in statistical physics and special relativity *The Study of Time* ed J T Fraser, F C Haber and G H Müller (Berlin: Springer) (see also *Stud. Gen.* **23** 1108 (1970))

6 Penrose R 1979 *General Relativity—An Einstein Centenary Survey* ed S W Hawking and W Israel (Cambridge: Cambridge University Press) p586

7 Penrose O and Percival I C 1962 The direction of time *Proc. Phys. Soc.* **79** 605

8 Prigogine I, George C, Henin F and Rosenfeld L 1973 A unified formulation of dynamics and thermodynamics *Chem. Scr.* **9** 5

9 Prigogine I 1980 *From Being to Becoming* (San Francisco: W H Freeman)

10 Jammer M 1974 *The Philosophy of Quantum Mechanics* (New York: Wiley)

11 Misra B, Prigogine I and Courbage M 1979 Lyapounov variables: Entropy and measurement in quantum mechanics *Proc. Nat. Acad. Sci. USA* **76** 4768

12 Stapp H P 1972 The Copenhagen interpretation *Am. J. Phys.* **40** 1098; 1980 Locality and reality *Found. Phys.* **10** 767

13 D'Espagnat B 1979 The quantum theory and reality *Sci. Am.* **241** May issue

14 Selleri F and Tarozzi G 1981 Quantum mechanics, reality and separability *Rev. Nuovo Cim.* **4** (2)

15 Born M 1949 *Natural Philosophy of Cause and Chance* (Oxford: Clarendon Press)

16 Peres A 1974 Quantum measurements are reversible *Am. J. Phys.* **42** 886

17 Clauser J F and Shimony A 1978 Bell's theorem: experimental tests and implications *Rep. Prog. Phys.* **41** 1881

18 Scully M O, Shea R and McCullen J D 1978 State reduction in quantum mechanics: A calculational example *Phys. Rep.* **43** 485

19 Cantrell C D and Scully M O 1978 The EPR paradox revisited *Phys. Rep.* **43** 499

20 Bondi H 1962 The Halley Lecture, Physics and Cosmology *Observatory* **82** 133

21 Brillouin L 1962 *Science and Information Theory* (New York: Academic Press)

22 Chambadal P 1972 *Paradoxes of Physics* (London: Transworld Publishers)

23 Basano L 1980 Farewell to Tachyons? *Found. Phys.* **10** 937

24 Cohen M H *et al* 1977 Radio sources with superluminal velocities *Nature* **268** 405

25 Rowan-Robinson M 1977 Aether drift detected at last *Nature* **270** 9

26 Landsberg P T 1978 *Thermodynamics and Statistical Mechanics* (Oxford: Oxford University Press)

27 Hawking S W 1977 The quantum mechanics of black holes *Sci. Am.* **236** January issue

28 Sciama D W 1979 Black holes and fluctuations of quantum particles *Relativity, Quanta and Cosmology* ed M Pantaleo (New York: Johnson Reprint Corporation)

29 Landsberg P T and Tranah D 1980 Entropies need not be concave *Phys. Lett.* **78A** 219

30 Landsberg P T 1959 The entropy of a non-equilibrium ideal quantum gas *Proc. Phys. Soc.* **74** 486

31 Dirac P A M 1979 The large number hypothesis and the Einstein theory of gravitation *Proc. R. Soc.* A **365** 19

32 Wesson P S 1978 *Cosmology and Geophysics* (Bristol: Adam Hilger)

33 Tolman R C 1934 *Relativity, Thermodynamics and Cosmology* (Oxford: Clarendon Press)

34 Gold T 1962 The arrow of time *Am. J. Phys.* **30** 403

35 Schumacher D L 1964 The direction of time and the equivalence of 'expanding' and 'contracting' world models *Proc. Camb. Phil. Soc.* **60** 575

36 Schmidt H 1966 Model of an oscillating cosmos which rejuvenates during contraction *J. Math. Phys.* **7** 494

37 Hawking S W and Israel W (ed) 1979 *General Relativity—An Einstein Centenary Survey* (Cambridge: Cambridge University Press). See particularly the articles by B Carter, G W Gibbons, S W Hawking and R Penrose (an extract from which appears here as paper 11).

38 Barrow J D and Silk J 1980 The structure of the early universe *Sci. Am.* **242** April issue

39 Hawking S W 1976 Breakdown of predictability in gravitational collapse *Phys. Rev.* **D14** 2460

40 Nicolis G and Prigogine I 1977 *Self-organisation in Non-equilibrium Systems* (New York: Wiley)

41 Landsberg P T 1980 Stability and dissipation: non-equilibrium phase transitions in semiconductors *Eur. J. Phys.* **1** 31

42 Landsberg P T and Evans D A 1977 *Mathematical Cosmology: An Introduction* (Oxford: Oxford University Press)

43 Hawking S W and Ellis G F R 1973 *The Large-scale Structure of Space–Time* (Cambridge: Cambridge University Press)

44 Tipler F J 1978 Energy conditions and space–time *Phys. Rev.* **D17** 2521

45 Jeans J H 1930 *The Mysterious Universe* (Cambridge: Cambridge University Press)

Minor Comments on
some of the Reprinted Papers

G N Lewis: The Symmetry of Time in Physics
Reference 3 refers to the Proceedings of the American Academy of Arts and Sciences of 1912 (not 1921). The law of entire equilibrium is now called the principle of detailed balance, and the mutuality principle is now called the principle of microscopic reversibility. It is easy to see that these two laws imply the law (3) of equiprobability in the paper.

E Schrödinger: Irreversibility
Schrödinger fails to point out in his Section 2 that the Poincaré recurrences can be removed by taking a sufficiently large system.

For his dialogue in Section 3 imagine the 'counters' shown in figure 5,

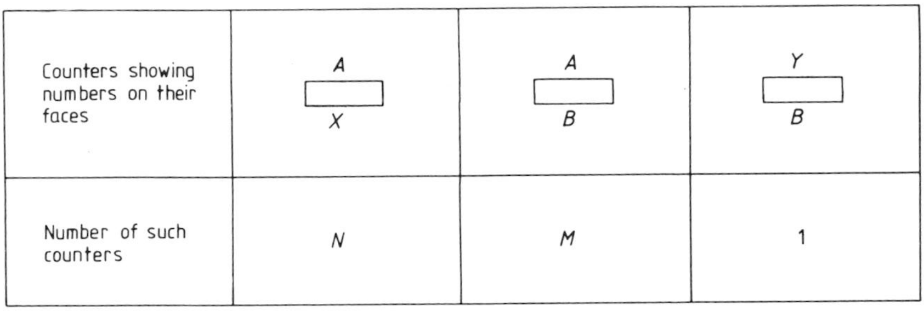

Counters showing numbers on their faces	$\begin{array}{c} A \\ \boxed{} \\ X \end{array}$	$\begin{array}{c} A \\ \boxed{} \\ B \end{array}$	$\begin{array}{c} Y \\ \boxed{} \\ B \end{array}$
Number of such counters	N	M	1

Figure 5. The counters for section 3 of Schrödinger's paper.

there being say $N \sim 10000$ counters of the first kind and $M \sim 100$ of the second. In Schrödinger's example $N = 81$, $M = 9$, $A = 3$, $X = 4$, $B = 2$ and $Y = 1$ and the 2 came up on the toss which made the chance of a 3 on the other side of the counter $M/(M + 1)$ or $100/101 \sim 99\%$ with our numbers here. This is a fairly safe guess, particularly if M is made even larger. So the dialogue is presumably meant to show that a reasonably safe inference can be made from probabilistic data. This matter is, however, not made very clear.

P Morrison: Time's Arrow and External Perturbations
The calculation on p. 58 is not quite clear and can be omitted.

F Hoyle: The Asymmetry of Time
The problem of the direction of time for radiation is also considered by G N Lewis.

P A M Dirac: New Ideas of Space and Time
There are a number of papers by I I Shapiro and his colleagues. Dirac presumably refers to the papers on the fourth test of general relativity which may be found in *Phys. Rev. Lett.* **20** 1265 (1968). See also *Phys. Rev. Lett.* **26** 1132 (1971).

P T Landsberg: Thermodynamics, Cosmology and the Physical Constants
Four quotations in the appendix come from papers reprinted in this volume—those by Aharony, Landsberg (A Matter of Time), Morrison and Hoyle.

A Aharony: Time Reversal Symmetry Violation and the H-theorem
Note that reference 5 in this paper is Morrison's paper reprinted in this volume.

Y Ne'eman: Time Reversal Asymmetry at the Fundamental Level—and its Reflection on the Problem of the Arrow of Time
Note that references 6 and 12 in this paper are Morrison's and Aharony's papers, respectively, reprinted in this volume.

A J Leggett: The 'Arrow of Time' and Quantum Mechanics
It may help to give a subdivision of this article.
1. The problem of the arrow of time and its relation to causality (p. 149 to the top line of p. 150).
2. Low disorder: Fluctuation or intelligent activity? (p. 150 to line 7 of p. 151).
3. Cosmological time asymmetry (first complete paragraph on p. 151).
4. Human agency 1: Time reversed beings, precognition (p. 151 to line 8 of p. 152).
5. Human agency 2: Measurement problem in quantum mechanics (p. 152 to line 10 from bottom of p. 153).
6. Hidden variables and Bell's theorem (p. 153 to line 4 from bottom of p. 154).
7. Violations of a uniform time sense (p. 154 to end of p. 155).

P C W Davies: Black Hole Thermodynamics and Time Asymmetry
The remark on p. 184 that ergodicity is expected for non-gravitating systems needs additions to become technically correct.

R Penrose: Singularities and Time-asymmetry
The sections included here are extracts from a longer chapter hence the occasional references to other sections. The extracts presented here can, however, be well understood without reference to the remainder of the chapter.

Part A

Irreversibility

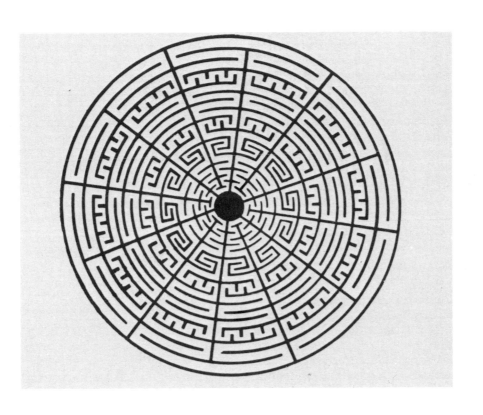

III. *An irreversible process.* A 100-gradation incense seal. When there was a drought in China in AD1073 it was not practicable to use the usual water clocks or clepsydrae for time measurement. A clock was invented in which powdered incense was allowed to burn along marked segments so that the passage of time was noted both visually and by the aroma. The negative, female, waning yin hours referred to the periods when the trail was inwards. At the end of the process it becomes yang, when the track leads outwards. This is associated with the strong, male tendency. (According to Chinese philosophers a purpose of life is to achieve harmony of the yin and yang.) The shape of the clock is reminiscent of a seal and it is thus called an incense seal. Surviving examples are made of metal.

> The incense stick has burnt to ash
> The water clock is stilled
> The midnight breeze blows sharply by
> And all around is chilled.

. . . Old Chinese poem, from A M Earle, *Sundials and Roses of Yesterday* (New York, 1902) p.54.

THE SYMMETRY OF TIME IN PHYSICS[*]

By Professor GILBERT N. LEWIS

UNIVERSITY OF CALIFORNIA

A FEW years ago I presented[1] the outline of a theory of light which required a radical change in our ideas of temporal causality. Instead of assuming the time-honored unidirectional causality, in which cause inevitably precedes effect, it proved necessary to assume that the present phenomena of a system are determined no more by the past states of the system than by its future states. Several recent developments in physics make this assumption seem less startling now than then; indeed I am fully convinced that there is no other way in which we can account for the known phenomena of light. Moreover, new discoveries in wave mechanics indicate that any conclusions concerning the emission of light must be extended to the emission of every kind of material particle.

By such considerations I was led, in "The Anatomy of Science," to examine with some care the meaning of time, as the word is used in physical science. It often happens that a common concept of daily life may profitably be simplified or refined when it is to be employed in a single branch of science. In studying the vastly complex phenomena of nature, as they come to us through our sense impressions, we could make little headway did we not segregate and idealize certain groups of like phenomena for the purpose of special study. Such segregations define the several branches of science, of which one of the most highly specialized and idealized is physics. Only a few types of phenomena are included within its bounds, and in its study we consciously abstain from employing many of our commonest ideas, such as purpose, goodness, beauty. In the physical sciences a statue of Praxiteles is a certain mass of crystalline calcium carbonate; the shape may or may not be mentioned. It was the scientific arrogance of a previous age that called a law of physics a law of nature. To speak so is to forget the bounds that we have ourselves established.

* Address given on the occasion of the presentation of the gold medal of the Society of Arts and Sciences, New York, April 17, 1930.
1 Nature, 117: 236, 1926.

It is therefore evident that such notions as those of time and space may be given a simpler significance when we are dealing with a single science than when we are concerned with the complexities of natural occurrences in general. Our common idea of time is notably unidirectional, but this is largely due to the phenomena of consciousness and memory. Was Newton right in deliberately introducing into physics this common idea of the *flow* of time? Surely in one great branch of physics which we owe to his genius, the mechanics of conservative systems, it has long been recognized that there is need for nothing more than the simple idea of symmetrical time, which makes no distinction between past and future.

These two ideas of time, the unidirectional and the symmetrical, I have for brevity called "one-way" and "two-way" time. In going from the very simple science of mechanics to the very complex science of psychology, we must change from two-way to one-way time. It is important to inquire where this transition comes, and whether two-way time suffices for some parts of physics while one-way time is needed for the remainder.

The thesis that I announced earlier, and now wish to elaborate, is that throughout the sciences of physics and chemistry, symmetrical or two-way time everywhere suffices. As a philosophic speculation this view has received some attention, but I shall be much disappointed if it can not also be accepted as the statement of a law of physics, of exceptional scope and power, directly applicable to the solution of many classical and modern problems of physics.

Let us therefore review the several great branches of physics in the light of this thesis of symmetrical time. These branches are mechanics, thermodynamics, theory of radiation and electromagnetics. We shall see that nearly everywhere the physicist has purged from his science the use of one-way time, as though aware that this idea introduces an anthropomorphic element, alien to the ideals of physics. Nevertheless, in several important cases unidirectional time and unidirectional causality have been invoked, but always, as we shall proceed to show, in support of some false doctrine.

MECHANICS

Mechanics includes the still more limited science of kinematics. For a century or more there have been attempts,[2] culminating in the brilliant work of Minkowski, to make kinematics a branch of geometry. It was the hope, now fulfilled, that time could be combined with space into a four-dimensional manifold, of which the geometry should reproduce the science of kinematics.

[2] *E.g.*, Fechner (Kleine Schriften), "Der Raum hat vier Dimensionen."

To lighten the discussion, let us imagine one of these precursors of Minkowski, whom we may call Dr. X. In one of his note books we might read, "If this geometrical view of kinematics is correct there must be no distinction of past and future. It would be absurd in Euclidean geometry to prove a theorem by means of a diagram, and then to claim that the theorem becomes invalid if the diagram is turned upside down. Likewise there is no up or down in the four-dimensional geometry of kinematics."

It was also the belief of Dr. X that the rest of mechanics could in turn be identified with a still more comprehensive geometry, and it seemed to him that this view received some corroboration in the fact that the mechanics of conservative systems requires no dissymmetry of time. All the equations of mechanics are equally valid when t is replaced by $-t$. The chance of error is the same in calculating an eclipse of a thousand years ago or of a thousand years hence.

Becoming even bolder, this eager speculator hoped that not only mechanics but all physics might eventually be reduced to a geometry. He wrote, "If this belief be correct, Newton's idea of a flow of time has no place in physics. Until I see strong evidence to the contrary, I shall maintain this to be a basic law of physics, that all rules which are obtained from a study of physical processes hold with equal validity if these processes are reversed in time. Every equation and every explanation used in physics must be compatible with the symmetry of time. Thus we can no longer regard effect as subsequent to cause. If we think of the present as pushed into existence by the past, we must in precisely the same sense think of it as pulled into existence by the future."

THERMODYNAMICS

The second law of thermodynamics was a source of uneasiness to Dr. X. Recognizing the importance of its consequences, he still objected to the statement of Clausius, namely, that in any system left to itself the entropy increases steadily toward a maximum. This statement is in direct defiance to the law of the symmetry of time. Therefore to Dr. X it was a great satisfaction to read in a paper of Willard Gibbs that "the impossibility of an uncompensated decrease of entropy seems to be reduced to an improbability"; and later to follow the development of this thesis by Boltzmann until near the end of the famous lectures on "Gastheorie" he found Boltzmann saying, "Hence, for the universe, both directions of time are indistinguishable, as in space there is no up or down."

Boltzmann's qualifications of this statement seemed unnecessary to Dr. X, who now definitely included thermodynamics among those branches of physics which require symmetrical time only. In his note book we read, "The statistical interpretation of thermodynamics offered by Gibbs and Boltzmann affords for the first time an understanding of entropy. The process irreversible in time does not exist. This corollary of the law of symmetry in time itself leads to further important consequences. Thence we may prove to those who are still skeptical the atomic structure of matter, as follows: if we imagine two continuous media to diffuse into one another, such a diffusion would in principle be a phenomenon which by no physical means could be reversed, but if two streams composed of discrete particles should diffuse, then, although it might be a matter of great difficulty to recapture the particles and restore each to its own kind, yet in principle the process is reversible and indeed, according to Boltzmann, the separation will occur spontaneously if the system be left to itself for a sufficiently long period."

Dr. X adds a remark of much subtlety. "While we recognize the particulate nature of matter, we allow each particle to have a position and a velocity chosen from a whole continuum of possible values. Thus while we claim that an isolated system repeatedly returns nearly to its initial condition, we can not say that it returns exactly to that condition. If we start with a number of molecules all moving in precisely the same direction, we can not claim that after some disturbance they ever again move quite parallel to one another. This implies a sort of irreversibility, and while I am not sure that it is a contradiction to symmetrical time, I confess that I should be better satisfied if we could claim the exact recurrence of an initial state."

It is a pity that Dr. X did not live to see the universal acceptance of quantum theory, which assigns to an isolated system not an infinite continuum of states, but a finite number of discrete states. Thus every particular state exactly recurs within finite time. This modern picture is far simpler than that of Boltzmann, especially as we are going to see that each particular state occurs as often as every other. Hence molecular statistics furnishes quite elementary problems in the theory of probability, like the tossing of coins or the shuffling of cards.

In the main, however, the problems of thermodynamics to-day are not far different from those discussed by Boltzmann and Dr. X. In the note book of the latter we read, "The earth is constantly receiving energy from the sun, and in consequence water is continuously flowing over Niagara Falls, but these descriptive statements can not be called laws of physics. When we turn to the processes studied in the laboratory we find that when a hot and cold body are brought together, it is almost certain that the two temperatures will become equalized until no

discernible difference remains. If we mix two mutually soluble liquids, we may expect the concentration to become nearly uniform. I have learned that it is possible to perform an operation upon the brains of mice so that they respond to no external stimuli, but can still run aimlessly about. If a large number of these mice are placed in one end of a box, that end is now heavier than the other; but this distinction rapidly disappears as the mice, in their random movements, cover with greater uniformity the bottom of the box, so that we may no longer discern any tendency of the box in one direction or the other. I claim that in all these cases there is no phenomenon irreversible in time, and indeed nothing more formidable occurs than in the proverbial case of a needle dropped into a haystack."

Before analyzing further these problems, we may consider a very interesting discussion of one-way time by Professor Eddington, in "The Nature of the Physical World." He arrives at a compromise, first by stating that one-way time does not occur in physics outside of thermodynamics, and then by reducing the principle of the increase of entropy from a "primary" to a "secondary" law, which does not prevent him, however, from deducing therefrom a "running down of the universe." To this compromise I can not agree. The first statement will be answered by the cases which will be discussed in the following sections, and the second can not be regarded as consistent with the new conception of thermodynamics.

We must be cautious about extending to the whole cosmos the rules which we have obtained from limited experiments in our small laboratories. The chance of obtaining valid results from such an extrapolation is very small, but it can be made in a purely formal way. If the universe is finite, as is now frequently supposed, then the formal application of our existing ideas of thermodynamics and statistics leads directly to the following statement: The precise present state of the universe has occurred in the past and will recur in the future, and in each case within finite time. Whether the universe actually is running down is, of course, another matter. All we can say is that such an assumption obtains no support from thermodynamics.

Let us, however, turn from the behavior of the universe, about which we know almost nothing, to the phenomena of the laboratory, about which we know a little more. Even in this limited domain it is going to be difficult enough to persuade ourselves that such a phenomenon as an explosion is wholly compatible with the thesis of symmetrical time. If a statement runs counter to a fixed habit of thought which has become nearly instinctive, it may be accepted by many, but believed by few. The use of one-way time has become second nature to us, and to oust from the mind all its implications, even when we set ourselves to do so, is no easy task. Nevertheless, perhaps we can make this task easier if we dig up by the roots and examine with all care this thing that we call an irreversible process.

We must begin by guarding against two human frailties—the feeling that there is some real distinction between familiar and unfamiliar things, and the fear of large numbers. Let us illustrate by means of a pack of cards, and at first a very small pack, say the ace, two, three and four of spades. There are twenty-four possible distributions, such as 1, 2, 3, 4; 4, 2, 1, 3; 4, 3, 2, 1, and so on. Of these the first and third are a little more easily described and remembered than the other twenty-two, which for this reason we call nondescript, but this is only a question of familiarity. A whist player would think of the arrangement, 1, 4, 3, 2, and there is no one of the twenty-four arrangements which might not be particularly significant in some other card game.

If our attention has been drawn to one particular distribution, we remark upon it when it results from random shuffling; but on the average each distribution, whether or not it has been favored by our attention, will turn up once in twenty-four times. If we now take a pack of fifty-two cards, the familiar, or easily described, distributions are relatively rare compared with all the nondescript arrangements, and if random shuffling should give the exact distribution of the pack as it comes from the manufacturer it would seem almost a miracle; yet we can say with the same certainty as before that any one particular distribution will, on the average, occur once in 52! times. The rules of arithmetic are the same for large numbers as for small.

There is no such thing as a well-shuffled pack, except with reference to certain familiar sequences. If the distribution does not closely resemble some familiar sequence, we might call it well shuffled. There is, however, another sense in which the idea of shuffling is of fundamental significance. If we examine a particular distribution and remember the sequence of the cards, afterwards the pack is said to be well shuffled when our remembrance of the previous distribution no longer aids us in guessing what a new distribution will be. This distinction between a known distribution and an entirely unknown one will prove to be fundamental in our study of the corresponding problems of thermodynamics.

Turning now to the irreversible thermodynamic process, we shall choose an illustration which is not quite so complicated as an explosion, but involves all essentials. A chemist has spent days in preparing a flask of nearly pure alcohol. This he places in a water bath, and then by accident the flask overturns and the alcohol diffuses through the water. His disappointment is in no way allayed by the fact that none of his material is really lost, nor by the belief that ultimately the molecules of alcohol will accidentally come together to give once more a nearly pure sample. That the chemist would be obliged to wait an unconscionable time for this chance restoration must be given no weight. If it occurred every ten minutes, the principle would be the same. It would still be necessary for him to devise rapid analytical methods to ascertain just when the fortunate event occurred. There is no question but that the accident has involved an element of *loss* which typifies the irreversible process (which is also spoken of as a process of dissipation, or degradation), but we shall see that this loss in no way implies a dissymmetry of time, nor indeed that it has any temporal implications whatever.

Without losing any of the characteristics of the reversible process, we may next examine one of the simplest of systems. Suppose that we have a cylinder closed at each end, and with a middle wall provided with a shutter. In this cylinder are one molecule each of three different gases, A, B and C, and the cylinder is in a thermostat at temperature T. In dealing with the individual molecules we are perhaps arrogating to ourselves the privileges of Maxwell's demon; but in recent years, if I may say so without offense, physicists have become demons.

Regarding each molecule, we shall ask only whether it is in the right or the left half of the cylinder. Obviously eight distributions are possible, such as A and B on the left and C on the right; or B on the left and A and C on the right. According to our ordinary assumptions, each of these distributions is equally probable, or, in other words, the system averages to be in each distribution one eighth of the time. Moreover, each of the eight distributions can be easily described and remembered, so that we are not troubled by a large number of nondescript states. Each distribution occurs over and over, but in no particular order, and in these occurrences there is no trace of dissymmetry with respect to time—there is no "running-down" process here.

Yet we may have a typical irreversible process. Suppose that the shutter is closed so as to trap a particular distribution, say all three molecules on the left. We become familiar with this one distribution and wish to study it further, but accidentally the shutter is opened, and instead of the one distribution, we have all eight succeeding one another in a random way. This is a complete analogy to the overturn of the flask of alcohol. If we desire once more to obtain and keep the one distribution in which all the molecules are on the left-hand side of the cylinder, we may exercise our prerogatives as Maxwell demons by closing the shutter from time to time and determining by spectroscopic means or otherwise which distribution is trapped. In about eight trials we shall obtain the desired result. Unless, however, there is in sentient beings the power to defy the second law of thermodynamics, we shall find that this method of obtaining the desired distribution requires at least as much work as the old-fashioned thermodynamical method of forcing the system into the particular distribution without the aid of demoniacal devices. This classical method consists in slowly pushing a piston from the extreme right of the cylinder as far as the middle wall. In this typical reversible process the work required to overcome the pressure of the three molecules is $3\ k\ T\ \ln\ 2 = k\ T\ \ln\ 8$. At the same time the entropy of the gas is diminished by $3\ k\ \ln\ 2$.

If we wish to obtain any other one of the particular distributions, from the general distribution, the same amount of work is required. Suppose the particular distribution desired is B on the left, A and C on the right. At the extreme left we have a piston permeable only to B, and at the extreme right a piston permeable only to A and C, and these pistons are moved slowly to the middle wall. We thus obtain the given distribution, and the sum of the work done upon the two pistons is $3\ k\ T\ \ln\ 2$. In every case, in passing from the general distribution to a particular known distribution, the gas loses entropy in the amount $3\ k\ \ln\ 2$. All these processes are completely reversible. If we start with any known distribution and let the proper pistons move outward from the center to the ends of the cylinder, we obtain the general distribution, the system does work in the amount $3\ k\ T\ \ln\ 2$, and the entropy of the gas increases by $3\ k\ \ln\ 2$.

The entropy of the general unknown distribution is greater than the entropy of any one known distribution by $3\ k\ \ln\ 2$. This, therefore, is the increase in entropy in the irreversible process which occurs when, after trapping any one *known* distribution, we open the shutter. It is evident, however, that the mere trapping of one distribution makes no change in the entropy, for the shutter may be made as frictionless as we please, and the mere act of opening or closing it will not change the entropy of the system. If we start with the shutter open, with all the eight distributions occurring one after another, and then close the shutter, the system is trapped in one distribution, but there is no change of entropy.

Whence we have now reached our most important conclusion. The increase in entropy comes when a *known* distribution goes over into an *unknown* distribution. The loss, which is characteristic of an irreversible process, is *loss of information*. In the simplest case, if we have one molecule which must be in one of two flasks, the entropy becomes less by $k\ \ln\ 2$, if we know which is the flask in which the molecule is trapped.

Gain in entropy always means loss of information, and nothing more. It is a subjective concept, but we can express it in its least subjective form, as follows: If, on a page, we read the description of a physicochemical system, together with certain data which help to specify the system, the entropy of the system is determined by these specifications. If any of the essential data are erased, the entropy becomes greater; if any essential data are added, the entropy becomes less. Nothing further is needed to show that the irreversible process neither implies one-way time, nor has any other temporal implications. Time is not one of the variables of pure thermodynamics.

THE THEORY OF RADIATION

The laws of optics are entirely symmetrical with respect to the emission and absorption of light. The whole science of optics leaves nothing to be desired with respect to symmetry in time. When time is considered reversed, the emitting and absorbing objects merely exchange rôles, but the optical laws remain unchanged. On the other hand, the physical theories concerning the radiation from a particle, which were for a long time current, introduced the idea of one-way time in a notable manner. Let us quote once more from the note book of Dr. X.

"It has always been conceived that a particle which has been set in vibration, perhaps by collision with another particle, dissipates its energy in a continuous expanding spherical shell, every part of which moves steadily out into space until it meets an absorbing body. Since the energy all comes from the vibrational energy of one particle, the whole is regarded as a unitary process, although those parts of the shell of energy which meet neighboring objects may be absorbed within a very small fraction of a second, while other portions may travel years before they meet an absorbing object. The exact physical reversal of such a process is quite unthinkable. We should be obliged to imagine some prearrangement whereby each of a number of bodies far and near would, at the appropriate time and in the right direction, send out its quota of energy, all of which in the neighborhood of the absorbing particle would coalesce into a continuous spherical shell. As a rare event it might by chance occur that something approximating to this picture would be observed, but in no case could an exact reversal of the assumed process of radiation be found.

"The emission of a continuous spherical shell of

energy is essentially irreversible in time, like the diffusion of continuous media which we have previously discussed. But we shall still be in trouble even if we assume that the energy radiating from a particle does not spread as a continuous shell, but goes to a limited number of other particles. Assume that a central particle emits energy to a number of other particles, and that its oscillations are damped as it loses energy, according to some simple law. The amounts of energy received by the several particles and the time of receipt are assumed to be causally connected, since the energy all flows from the one simply damped central atom.

"The exact temporal reverse either of this process or of this explanation is absurd. It would be necessary to imagine a central atom which receives energy from other atoms in such amounts and at such times as to increase continuously the oscillations of the central atom, according to a law exactly opposite to the law of damping; and we should be obliged to explain this phenomenon by saying that the amounts of energy emitted by the several particles, and the time of their emission, are all causally connected, by the fact that the energy is all to be received, and in a specified manner, by the central atom.

"After many considerations of this character I have come to the conclusion that the only process of radiation which can be harmonized with the symmetry of time is a process in which a single emitting particle at any one time sends its energy to only one receiving particle."

I think we may now agree that Dr. X was right and that if we are to assume the principle of symmetry in time, we are led irresistibly to a theory of radiation which has some of the characteristics of Einstein's theory of the light quantum. In particular we can not admit the possibility, now occasionally assumed, that in a single quantum process an atom may emit two photons to two separate atoms. Furthermore, in this theory of radiation we must assign to the emitting and to the absorbing atom equal and coordinate rôles with respect to the act of transmission of light, as I proposed in my former paper.

Electromagnetics

Our friend Dr. X was eminently satisfied with the development of the theory of electricity and magnetism. He saw that the equations of Maxwell would be equally valid if time were reversed, and was therefore bewildered by Maxwell's deductions from these equations of a theory of radiation which, from his point of view, had all the faults of previous theories. How was it possible to obtain the old one-way theory of radiation from equations which involve nothing but symmetrical time? He did not see the full answer

to this question until the development of the theory of *retarded potentials*. Then he realized that in the mathematical treatment of the problem two symmetrical solutions arose, of which one was arbitrarily discarded because it seemed inconsistent with common notions of causality.

In the whole history of physics this is the most remarkable example of the suppression by physicists of some of the consequences of their own equations, because they were not in accord with the old theory of unidirectional causality. We shall therefore attempt to analyze this problem, especially as this abstruse subject may be rendered quite simple by geometrical methods.

For this purpose Professor Wilson and I invented[3] the geometrical vector field. A vector field need not involve physical quantities such as momentum or force, but may be purely geometrical; for example, a line in space may serve to define any number of such vector fields. Thus we may, from every point in space, draw a vector along the perpendicular to the line, and with a magnitude proportional to the distance from the line.

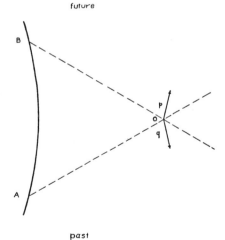

FIG. 1

In the space-time of relativity the geometry is characterized by the *singular* lines passing through every point, which are interpreted physically as light-paths. In the accompanying two-dimensional diagram they are represented for the point O by the dotted lines. In this geometry we set up the following vector field. Given any curve of the type AB, as the source of the field, then at any point O through which pass the singular lines OA and OB, the vector *p* is drawn

[3] *Proc. Amer. Acad.*, 48: 389, 1921.

upwards, parallel to the tangent of the curve at A, and with a magnitude which is the reciprocal of the distance from O to that tangent. The geometrical field thus set up has very remarkable properties. Our rules determine the variation of p in the neighborhood of O, and thus all its derivatives. The equations obtained are identical with all the complicated equations of the electromagnetic field produced by a moving and accelerated charge. This parallelism becomes an identity when we consider the curve AB as the locus in space-time of an electrical charge, and multiply the vector p by the magnitude of the charge. This vector, when projected upon a chosen time-axis, and upon the corresponding space, gives the scalar and the vector parts of the so-called retarded potential. It was so named because the influence of the charge at A was supposed to travel outward with the velocity of light and reach the point O at a later time.

Returning, however, to our geometry, we see that since there is no distinction between up and down, it is quite impossible to define the vector p without at the same time defining the vector q, which is drawn downward parallel to the curve at B. The projections of the vector q (multiplied by the charge) are quantities which have occasionally been studied under the name of *advanced potentials*. If they alone had been employed, the retarded potentials being discarded, we should have had an electromagnetic theory of light which would have been in equally good agreement with experimental facts, but in the interpretation of which we should have been obliged to regard the absorbing particle as the active agent, sucking in energy from all parts of space, in a spherical shell which contracts with the velocity of light.

As we now know, neither of these two electromagnetic theories is correct, and they can be used only as analogues; but in using such analogues we must hereafter give equal and symmetrical consideration to the retarded and to the advanced potentials; which means that in any theory of light we must consider the emitting and receiving agents as of coordinate importance. Thus, for example, if we wish to consider the probability that an atom X will emit a photon to an atom Y, and for this purpose imagine a virtual field produced by the particle X because of something analogous to its retarded potential, we must at the same time consider the particle Y as the seat of another virtual field of the advanced instead of the retarded type, and these two fields must be combined in a symmetrical manner to give the probability in question.

In recent months attempts have been made to extend quantum mechanics to the electromagnetic field, and here again the retarded potentials have been employed. We may safely predict that such attempts will not fully succeed until the retarded and advanced potentials are used simultaneously and symmetrically.

APPLICATIONS TO NEW PROBLEMS

We have seen that if science long ago had accepted the principle of symmetry in time, it would have eliminated the idea of unidirectional causality which has led to so many of the errors of classical physics. From this principle could have been deduced the atomic structure of matter and the newer thermodynamics. By its aid the flaws in the older theories of radiation and in the electromagnetic theory would have been seen. Moreover, the idea that light passes only from one particle to one particle, and that in this process the emitting and receiving atoms play coordinate parts, was directly derivable from the law of the symmetry of time. Let us now see whether there are new and unsettled problems which may be similarly solved.

Of the utmost importance to chemistry is the problem of reaction rates. Knowing only the state of equilibrium in a chemical reaction we know nothing of the rates of the individual reactions; but if we know the laws governing the rate of a reaction in both directions we may calculate the conditions of equilibrium. We now find that the complex problem of reaction rates may be reduced to the simpler problem of the transition probabilities between two elementary states. It will be a long time before many of these transition probabilities or intrinsic reaction rates can be calculated, but we shall see that there is one fundamental law that governs them.

When we say that we have chemical or thermal equilibrium, we mean that the average amount of each chemical substance (and also the number of particles of each species lying within a specified region, such as a region of energy), on the average remains constant. If, for example, substances A, B and C can change one into another the amount of each of these substances in equilibrium will not change, but thermodynamics alone tells us nothing of the paths by which they may go. For example, we might assume rapid processes from A to B, B to C and C to A, and slow processes for the reverse direction, A to C, C to B and B to A. There has, however, been a growing tendency to regard as impossible all such "cyclic equilibria." The principle was used in a limited way by Boltzmann, and was taken over in the quantum theory of the kinetics of gases. There was, however, a few years ago, a general disinclination to extend the principle to systems involving radiation. I believe I was the first to set up this principle[4] as a universal law in all physics and chemistry, applicable not only to chemical and physical proc-

[4] *Proc. Nat. Acad. Sci.*, 11: 179, 1925.

esses involving material substances, but also to processes involving light. I called it the principle of entire equilibrium. It has also been called the principle of microscopic reversibility and the principle of detailed balancing. It states that in equilibrium the rate of change along every detailed path is equal to the reverse rate.

Led to the formulation of this law by the idea of symmetry in time, which I was then beginning to develop, I remarked, "The law of entire equilibrium might have been called the law of reversibility to the last detail. If we should consider any one of the elementary processes which are occurring in a system at equilibrium, and could, let us say, obtain a moving-picture film for such a process, then this film reeled backward would present an equally accurate picture of a reverse process which is also occurring in the system and with equal frequency. Therefore in any system at equilibrium, time must lose the unidirectional character which plays so important a part in the development of the time concept. In a state of equilibrium there is no essential difference between backward and forward direction in time, or, in other words, there is complete symmetry with respect to past and future."

Indeed, we can readily see that any cyclic equilibrium would mean dissymmetry in time, for, suppose that in the case cited above we could say that the process occurring followed chiefly the route ABCAB . . . , then if time were reversed, we should obtain the opposite rule, namely that the main route would be ACBAC . . . .

Consider for any system the completely detailed quantum states designated as a, b, c . . . , the law of entire equilibrium states that the system changes from a to b as often as from b to a. Now the chance of a transition from a to b is proportional to the probability, p_a, of finding the system in a multiplied by an intrinsic probability, φ_{ab}, that when the system is in a, it will go over in a given time to b. The law of entire equilibrium therefore states that

$$p_a\varphi_{ab} = p_b\varphi_{ba} \qquad (1)$$

Let us now examine these intrinsic probabilities, φ_{ab} and φ_{ba}. There seems at first sight nothing in the symmetry of time to restrict the values of these quantities. Supposing for the moment that these are the only two states, and assuming that, on the average, the system remains twice as long in the state a as in state b, the same would be true if time were reversed. A moving-picture representing the successive changes would look the same if it were run in either direction. However, science can not rest content with such a statement regarding the intrinsic probabilities; it

immediately inquires what physical quantities determine these probabilities.

According to the old idea of causality, the probability of a transition would be determined by the properties of the state which existed *before* the transition. In other words, the probability of the transition $a \rightarrow b$ would be some function of the properties of the state a, and the transition $b \rightarrow a$ would be the same function of the corresponding properties of the state b. Such a view is no longer permissible. If a transition depends upon the properties of the state preceding the transition, it must in equal measure depend upon the properties of the state following, so that φ_{ab} must be a symmetrical function of the properties of a and b. Since φ_{ba} must be taken as the same symmetrical function of the same properties, we obtain immediately the most fundamental law of physical and chemical processes,

$$\varphi_{ab} = \varphi_{ba} \qquad (2)$$

This law stating the equality of direct and reverse transition probabilities has received no name, except in so far as it has occasionally been confused with the law stated in equation (1). We may call it the law of the mutuality of elementary processes, or, more simply, the mutuality principle. The name is intended to suggest the important fact that a transition in one direction and a transition in the opposite direction are not two physical entities, but one entity looked upon in two ways. Whatever we can say of one process, we can say of the other. We may think of a double arrow rather than of two arrows pointing in opposite directions. At present the law is best illustrated by some of the equations of quantum mechanics, such as the equations of Schrödinger in which transition probabilities are expressed as symmetrical functions of the "proper functions" of two states.

The law of mutuality holds for the elementary states, or, in other words, for the completely specified quantum states of a system. When a system is said to be in a condition which comprises a number of elementary states, the probability of a transition from one such condition to another depends not only upon the properties of the elementary states, but also upon their number. Thus, for example, if condition α comprises only one elementary state a and condition β comprises the two elementary states b and c, the probability of a transition α → β is the sum of φ_{ab} and φ_{ba}, but if the system is in the condition β the probability of the reverse transition is not $\varphi_{ba} + \varphi_{ca}$ but is less, owing to the fact that when the system is in condition β it is in state b *or* state c, but not in both.

Generalizing, we may say that when we are dealing with a complicated chemical reaction in which a condition α goes into, or proceeds from, a condition β, and if we find that the specific reaction velocity is greater in the direction α → β than in the direction β → α, it signifies that there are more elementary states comprised in the condition β than in the condition α.

By combining equations (1) and (2) we obtain a third law which is known as the equality of *a priori* probabilities and which, since it does not involve the element of time, we need discuss no further here. It is

$$p_a = p_b \tag{3}$$

These three laws, of which, in fact, the first and third are both derivable from the second, are the fundamental laws of quantum kinetics and quantum statistics.

They are at present the most important deductions from the law of the symmetry of time.

It is remarkable that so many positive conclusions result from the negative statement that physics requires no one-way time, but more important conclusions have been derived from the similar negative statements that we can not have a perpetual motion machine and that we can not determine absolute velocities. Whether the new law will be successful in leading to new and unexpected conclusions remains to be seen. At least, if accepted, it will warn us away from certain lines of thought which involve one-way time. There is at present in the study of quantum mechanics and in some interpretations of Heisenberg's uncertainty principle a tendency to introduce anew the idea of unidirectional causality. I feel convinced that this is a retrograde tendency which may introduce new errors into science.

45

12.

IRREVERSIBILITY.

By E. SCHRÖDINGER.

(From the Dublin Institute for Advanced Studies.)

[Read 23 MAY, 1949. Published 28 AUGUST, 1950.]

1.

IT may seem an audacity if one undertakes to proffer new arguments in respect of a question about which there has been for more than eighty years so much passionate controversy, some of the most eminent physicists and mathematicians siding differently or favouring opposite solutions — Boltzmann, Loschmidt, Zermelo, H. Poincaré, Ehrenfest, Einstein, J. von Neumann, Max Born, to name only those who come to me instantly.   But, to my mind, in this case, as in a few others, the "new doctrine" which sprang up in 1925/26 has obscured minds more than it has enlightened them.   It is sometimes believed that only quantum mechanics, or some processes of thought borrowed from it, give the final clue to the problem.   I wish to show here that this is wrong and that the solution given previously can be defended against the last objection that continues to be raised again and again.

This objection, in short, is this : a proof that a reversible model shows an irreversible behaviour, i.e. that it "nearly always" exhibits a temporal succession of observable states which it " almost never " passes through in the reverse order of time—such a proof needs must be at fault somewhere.   This consideration seems so absolutely irrefutable that it obtrudes itself to the most sagacious minds again and again as an irremovable stumbling block.   Of late[1] a new way out has been sought— it is the one I had in mind when I spoke of the borrowing of thought from the new doctrines.

The following is known and is universally agreed upon : the overwhelming majority of all those micro-states that would impress our crude senses as the same observable (= macro-) state do lead to identical, moreover to the actually observed *consequences*.   That seems fine.   What ails us is only, that we can equally well scan the *antecedents*.   And they are—again for an overwhelming majority—entirely wrong, inasmuch as

---

[1] Max Born, Natural Philosophy of Cause and Chance, Oxford, at the Clarendon Press, 1949.

the antecedents are the mirror image in time of the aforesaid consequences; it would thus appear that the system has reached its momentary state by an "anticipation" of its actual future history in reversed order. It is true that our two "overwhelming majorities" do not exactly coincide. In particular, among the first lot—that with correct consequences—there is a small subset which has correct antecedents as well. But the microstates of this subset are so rare, even a little rarer than those we are inclined to neglect for having not the right consequences. It is therefore hard to see how the overwhelming majority concerns us at all and what benefit for understanding the observed phenomena we could draw from its, as it were, 50% correct behaviour.

From this awkward situation Born, *l.c.*, if I understand him aright, proposes the following rescue. Since we do not know the actual microstate of the system, we must—and that is where the philosophical loan from quantum mechanics comes in—refrain from drawing inferences from it. We must draw conclusions by averaging over all the microstates that may equally well be at the back of the observed macrostate. That looks splendid. For, after what has been said before and is agreed upon by everybody, we thus arrive at a correct prediction of the system's future behaviour. But it would seem to me a rather crude way of killing off the undesirable inference with respect to the opposite direction in time, if one prohibited any conclusions concerning the past by saying that our observation of the system in that particular moment is in itself an irreversible process which cannot be used for drawing any inferences concerning the past. Yet I can see no other way of freeing the conclusions drawn, after averaging, from the unfortunate symmetry with respect to time that has bothered us before. Surely the system continues to exist and to behave, to undergo irreversible changes and to increase its entropy in the interval between two observations. The observations we might have made in between cannot be essential in determining its course. Moreover, the very spirit of quantum mechanics, combined with that of thermodynamics, forbids us even to *think* of such observations taking place, if—as has often to be assumed—the system is isolated from the rest of the world in the interval between the two observations.

2.

The problem before us here is *not* actually to derive irreversibility—say, the increase of entropy with time—from any kind of general or special reversible model. Not from a general one: for it is hardly possible to devise a model general enough not only to comprise all kinds of physical events but also to anticipate all changes the reversible theories of physics may undergo in future, and to be inviolable to any such change; I mean changes as we have experienced them when Newton's absolute notions of space and time had to yield to the Theory of Relativity or classical mechanics to quantum mechanics. Still less would it serve our purpose if we were only to refute the objections raised against some special model, e.g. Boltzmann's model of a gas, purporting to picture certain irreversible happenings.

Our scope is in some respect narrower, in other respect wider. I do not wish to derive irreversibility at all. I wish to reformulate the laws of phenomenological irreversibility, thus certain statements of thermodynamics, in such a way, that the logical contradiction *any* derivation of these laws from reversible models seems to involve is removed once and for ever.

The task is clearly outlined. No such derivation can avoid introducing right at the outset a time variable t. If the model—whether it be a visualizable model of the old style or just a system of equations and prescriptions as is nowadays favoured in some quarters—I say, if the model is reversible, any general behaviour you rightfully infer for increasing t, must also hold for decreasing t. In other words it must be an invariant of the transformation $t' = -t$. Hence our task is to formulate all statements about irreversibility in such a fashion that they are invariant to the said transformation. At first sight it would seem that phenomenological time can have nothing to do with the variable t. It could not be defined by t. And it could not be defined by $-t$. This is true. And if you unite these statements and say it can be defined neither as t, nor as $-t$, that is also true. We shall see however that it can be defined as "either t or $-t$".

The most usual way of enouncing the Second Law is to say that a system perfectly isolated from the rest of the world never decreases its entropy, and, apart from the exceptional case that it happens to be already in thermodynamical equilibrium, increases its entropy until thermodynamical equilibrium is reached.

No model that has ever been conceived behaves in this way. Left to itself for a sufficient time it will take on all possible states again and again. Its entropy decreases as often as it increases. What is true is that only in an infinitesimal fraction of all the time you will encounter the model in a state *not* perceptibly corresponding to thermodynamic equilibrium. Moreover *if* you encounter it in such a non-equilibrium state then—supposing you have made a record of its history—you will find that it has left a state close to equilibrium comparatively not long before and that it returns to one not long after. During the time intervals of ascent to the non-equilibrium state and of the return from it, the quantity corresponding to entropy in the model behaves, apart from imperceptible fluctuations, monotonically, it decreases during the first and increases during the second of these intervals.

All this is well known. Perhaps less well known is the following. If you know that during the period of ascent or during the period of return your system has separated into two systems isolated from each other (as may happen), these two systems will also have their entropy changing monotonically (apart from small fluctuations), decreasing or increasing, as the case may be, but both in the same direction of time. While everybody will be prepared to grant this for the period of return to equilibrium some may be loath to accept my statement concerning the period of ascent, thus of decreasing entropy. To them I need only answer, that the two enouncements stand and fall together, since the model is supposed to be reversible. But perhaps it is well to tell the reason also in the customary jargon which calls the period of ascent an "infinitely improbable" one, the period of return one "following the ordinary laws of nature," necessary and unavoidable once the system has had the audacity to escape into this "frightfully improbable" state. (Actually, of course, since you know the system has escaped—and we had *chosen* such a rare moment—there is no longer anything improbable about it, it is just certain.) Generally speaking the periods of escape and the periods of return are exact time-mirror-images of each other. If you consider the period of escape as an extremely improbable one, and include the case of splitting mentioned above, well then you must tell yourself this: I *know* the system to have reached this very abnormally low value of the entropy (in fact I have waited until it did!). It is so infinitely less probable that the system should reach this state otherwise than by a monotonical decrease of the entropy during all the period in all its parts—even when separated—in a word otherwise than by a direct time-mirror-image of normal behaviour, that it is next to certain that it did follow this way. ("Infinitely less probable" has here the clear-cut meaning: only in an infinitesimal fraction of all the cases when a certain low value of the entropy has been reached would it have been reached in another fashion.)

3.

It is now quite obvious in what manner you have to reformulate the law of entropy—or for that matter all other irreversible statements—so that they be capable of being derived from reversible models. You must not speak of one isolated system but at least of two, which you may for the moment consider isolated from the rest of the world, but not always from each other. Envisage them for a time that ought not to be *too* long (but if it does not substantially exceed the time the universe has existed in its present form there will be no trouble). Let them be isolated from each other between the "moments" t_A and $t_B > t_A$ of that time variable t of which we spoke above, but in contact for $t < t_A$ and for $t > t_B$. Labelling the two systems by 1 and 2 and calling S_{1A} the entropy of system 1 at t_A, etc., the formulation of the entropy law I propose is

$$(S_{1B} - S_{1A})(S_{2B} - S_{2A}) \geqslant 0 ,$$

with the corrollary that, whenever an entropy difference is different from zero the change is (apart from imperceptible fluctuations) monotonical. If at least one of the differences is posifive, t is the time, if at least one is negative, $-t$ is the time, if they are both zero, this experiment has not succeeded in deciding the issue.

To get back to the ordinary formulation you may take S_1 to refer to the system under consideration, S_2 to the rest of the world. There is no danger of contradictory time definitions ever resulting from various experiments, since every system is in contact with the rest of the world when you observe it.

Once time—time's arrow—is settled in this manner it is no longer extraordinary to find that with respect to it friction, diffusion, viscosity and whatnot act in the fashion they do and not in the opposite fashion. In itself the latter is equally possible and even in a way equally probable with reversible models, since in every display of such phenomena the bodies involved find themselves at every moment in a microstate the time-mirror-image of which would lead to the whole process running backwards through all its previous stages in opposite order of time. But all these processes imply an entropy change and will therefore fit into our picture with the correct arrow, not spoil it by the wrong one.

It is hardly necessary to mention that our inequality is to be understood in the same approximate sense as the customary statement about the increase of entropy. If one of the entropy differences does not appreciably exceed the normal thermodynamic fluctuation of the system in question, the product is to be considered "practically zero." This might give rise to various objections, particularly when one of the two systems is so very big as "the rest of the world". If one feels uneasy about it, one may use a more cautious formulation. Instead of stating, as we did, that the product is greater than or equal to zero, it is quite sufficient to maintain that it is positive whenever both its factors appreciably exceed the normal thermodynamic fluctuations of the systems to which they respectively refer.

4.

I beg to be allowed to enhance this paper by a brief dialogue in which a physicist who refuses to use reversible models for representing irreversible events is likened to a prisoner, who is afraid of drawing a fairly safe conclusion, thereby missing the opportunity for ending his detention. The numbers on the counters represent observed entropy values. The numbers *of* counters (81, 9, 1) and their ratios ought to be astronomical. This is the story:

James was in prison with no hope of being released. One day the gaol-keeper came to his cell and said: I am in the position of offering you a chance for freedom. Would you accept it, if, in doing so, you ran the risk of losing your life, though only with odds of 1 in 10 against you?

— Certainly I would, said James, who is not a coward.

= Would you also accept it if the odds were equal for your either being executed or getting free?

— I would not, said James.

= Very well, said the gaol-keeper, you need not decide until the very last moment. What I am ordered to propose to you is a sort of gambling, for your life or for freedom. I have here an urn. I am putting into it 81 counters, each with a **3** on one side and a **4** on the other side. I am adding 9 counters with a **3** on one side and a **2** on the other side. And finally one counter with a **2** on one side and **1** on the other side. Shuffle them.

— I have. Now tell me what the game is going to be. You make me curious.

= Listen. You will draw one of the counters out of the urn at random and, without looking at it, toss it up into the air, then look at the side it shows. You are to guess what is on the other side. If you guess right, you are free, if you guess wrong, you'll be put to death—but you may refuse to guess, then you remain in gaol, with no danger to your life.

After thinking a while, James exclaimed:

—. Of course I accept.

= Be careful, said the gaol-keeper, reserve your decision until after tossing. You might be very unlucky in the counter you draw.

— How should that be, said James. If I see a **4** or a **1**, I am saved anyhow. If I see a **3**, I can pretty safely guess **4**, and, with a **2**, pretty safely guess **3**. That would be just the odds of 1 in 10 of which you spoke. I shall run that risk.

= Well, I hope you'll be lucky. But, mind you, you may still refuse to guess after having tossed. That is the inalienable rule of the game.

51

James drew and tossed and up came a **2**. He was on the point of opening his mouth to say: three—when the gaol-keeper violently put his hand on James' lips and said: Think, before you decide.

James was angry and worried. But soon the following occurred to him:

Well, I believe this counter to be one of the nine 2/3 counters (rather than the 2/1 counter, of which there is only one) and all my hope is set upon the correctness of this guess. But *if* it is correct, then, in the moment of tossing, there was an equal chance for the counter coming down the other way, showing me a **3**. Then the same mathematical principles which I am on the point of using with confidence would ruin me.

In fact, James realized that to a person who followed these mathematical principles the mere incident of drawing a 2/3 counter involved a 50% death-danger. This made him lose the courage to use those principles when they definitely seemed to point to such a dangerous counter. But he knew of no better principles and therefore definitely refused to guess and thus remained in prison.

I don't know was it mercy or cruelty that made the gaol-keeper snatch the counter at this moment and throw it back into the urn, so that nobody ever knew what was on the other side.

What do *you* believe?

Poor James went almost mad about it. When he realized that he had been fooled, he wrote into his diary:

Never be afraid of dangers that have gone by! It's those ahead that matter.

TIME'S ARROW AND EXTERNAL PERTURBATIONS

P. MORRISON

Physics Department, Massachusetts Institute of Technology

Cambridge, Massachusetts

(Received, June 28, 1965)

Willard Gibbs wrote:[1]

"Let us imagine a cylindrical mass of [continuous] liquid of which one sector of 90° is black and the rest white. Let it have a motion of rotation about the axis of the cylinder in which the angular velocity is a function of the distance from the axis. In the course of time the black and white parts would become drawn out into thin ribbons ... wound spirally about the axis. The thickness of these ribbons would diminish without limit, and the liquid would therefore tend toward a state of perfect mixture of the black and white portions. That is, in any given element of space, the proportion of the black and white would approach 1 : 3 as a limit. Yet after any finite time, the total volume would be divided into two parts, one of which would consist of the white liquid exclusively, and the other of the black exclusively."

It is from this *anschaulich* argument of Gibbs that the notion of coarse-graining in statistical mechanics can be held to flow. For it is now clear, as he himself puts it, that the uniformity of equilibrium, which is the result of the stirring in the liquid analogy, is conditional; given any degree of stirring, I can find full non-uniformity if I look very closely. But given any method of defining density (say, in phase space) with a finite cut-off to the information sought, any averaging or "coarse-graining", and the measurement of uniformity becomes certain. While there is a large and sophisticated literature on this problem * (which it were folly to claim to know) it remains probably the case that something equivalent to Gibb's coarse-graining process, whether *Stosszahlansatz* or random phase approximation [2], is an essential feature of all studies of the approach to equilibrium, of the

---

* *Note added in proof:* Very similar ideas have indeed been published by a number of authors! e.g., J. M. Blatt, Progr. Theoret. Phys. 22 (1959) 745.

arrow of time. There is a subjective element to this procedure which is a little disquieting. Are the subtle correlations still present in equilibrium, speaking strictly classically and in ideal cases, or are they not? Is the arrow of time then only an illusion? It is the purpose of this note to answer stoutly that the arrow is real, that is, not subjective, that it is not essentially cosmological, that it arises from an inescapable feature of all physical theory.

Let me begin with a concrete analogue [3]. Across the wall of my office there stretches five meters of computer output. A few hundred small rigid spheres, packed pretty closely into a flat box, have been followed through a couple of thousand collisions. To begin they are arranged in a regular square lattice. Then each ball is given two random velocity components $v_{x_i}(0)$, $v_{y_i}(0)$, though all move at the same speed, each starting to move from its lattice position. The collisions go on, and after what amounts to a few collisions per ball, the lattice has been stirred into randomness. The computer prints out "snapshots" of the configuration at our will. At a certain time t_R the motion is stopped, and the velocity components of each ball reflected, with $-v_{x_i}(t_R)$ for $v_{x_i}(t_R)$ and $-v_{y_i}(t_R)$ for $v_{y_i}(t_R)$. Now the motion retraces its wildly complicated path, and after the right number of collisions, plus a collisionless interval to retrace that precise time $t = 0$ before the first collision, the regular lattice has been marvelously restored. *But the reversibility is not certain.* It is dependent upon a knowing programmer, for the inescapable round-off errors coming from solving the equations of motion with only finite digital accuracy, in the field of rationals, so to speak, will always oppose reversibility, and often leaves the array with the same sort of chaos it had when the reversed motions began at t_R. Thus the computer has done the equivalent of what coarse-graining can do; it has introduced a subtle sort of noise, arising from finite knowledge, into the classical equations which assert infinite precision, but only in an unattainable analytical calculation. My main point is to add that every classical statement of the laws of motion of any system necessarily leaves out a small physical perturbation, some δH, which cannot in principle be included for finite systems, and which in fact is always amply large enough to prevent a complete retracing.

The argument is elementary, and at bottom not new. It is an inver-

IV. *When I have seen Time's fell hand defaced* . . . (*Shakespeare,* Sonnet 64). The Persistence of Memory, by Salvador Dali (1931). Dali often describes the subconscious world (e.g. hallucinations, the juxtaposition of unrelated objects) and this picture illustrates the malleability of solid forms and the transitory nature of time.

sion of that of Poincaré, who long ago put it that probability itself could be regarded as an illusion, in that the roulette wheel could in principle be taken as purely causal. It is only that the prediction of *rouge* is extraordinarily unstable to error of initial data! By extension, that prediction is also unstable to external perturbations. Therefore, a causal universe, classically without probability at all, becomes a statistical one whenever we consider systems in partial isolation from their context. For then the neglected interactions disturb the predictions of mechanics, and prevent us, say, from unstirring Gibbs's milk and ink. Whenever we choose to place the system boundaries, *something* remains outside which, in sufficient time and for suitably complex systems, will wreck the extraordinarily delicate correlations of position with velocity upon which reversibility, for example, depends. A simple estimate of the degree of sensitivity to external perturbations is the end part of our story. Note that only one system, the whole universe, could possibly exist without any external unknown perturbations. Any theory of a system less complete must allow their presence in some degree. If that degree is adequate, the system becomes irreversible, in spite of the reversibility of dynamics. The whole point is that the intricacy of the ribbons of black threading the phase space rapidly becomes so great for any system of many particles that even dynamically negligible, unshieldable, gravitational perturbations are competent to mix up the pattern. Of course time-reversing both system and perturbation would always work. But that means enlarging the system. For now the perturbation needs to be known, and must become part of the system. Still there remains some other disturbance outside. Only the whole universe can then escape, as it ought to escape, the requirements of the *Stosszahlansatz*.

Consider a system of many particles, say with f coordinates in phase space. It is located in the neighborhood of some mean $\bar{p}$ and $\bar{q}$, with a range in each: Δq, which represents the edge of its containing volume in coordinate space, and Δp, a measure of the r.m.s. momentum spread as well. It is enough to consider what collisions do to the p coordinates of the representative point in the hyperspace. At each effective collision, p coordinates move by an amount roughly equal to the typical measure of p spread, say Δp. After a time $T = N\tau_{coll}$, where N is the number of collisions per particle and τ_{coll} a mean collision time, the representative point has made a wildly complex path in the hyperspace. We may estimate the projected distance between successive crossings of some typical value of one p coordinate as $\Delta p/N^m$, where the power m of N may correspond to some sort of random walk. (Whether it is 1 or $\frac{1}{2}$, or any small number, makes no difference to our argument.) Now the volume of a typical little grain of momentum space which is missed by the trajectory is about $(\Delta p/N^m)^f$. If during a time T a neglected external force shifts the $\bar{p}$ value by an amount δp, the volume of momentum space held tangent between new and old trajectory amounts to about $(\delta p) \times (p^{(f/2-1)})$. When the empty volume is about equal to the unforeseen volume change, we may expect an error .n the trajectory which makes it entirely different from the prediction, at the scale required for prediction. Reversibility, for example, would be lost; only quasi-ergodic predictions would be secure. But this means $\delta p \sim \Delta p/Nf'$ ($f' \sim kf$, k is a number of order unity), and then $T_{\text{mixing}} \approx \tau_{\text{coll}}(\Delta p/\delta p)^{1/f'}$.

The true solution is so filigreed and braided that the slightest external effect soon shifts it by an amount characteristic of its own scale of detail. One may estimate that a gravitational force exerted by a falling apple a kilometer away over an arc of ten centimeters is ample to mix up the trajectory of a mole of normal gas, in a time of milliseconds. Admittedly this has been a wildly crude estimate, but I do not believe it is in substantial error. For a less complex system, the perturbation becomes of dwindling effectiveness; the solar system cannot be treated as reversible in the presence of galactic forces, but the earth-moon system is easily managed to high accuracy. A few molecules would work equally well simply held in a box.

Gibbs and many followers have emphasized the importance of a *large* strongly-coupled thermostat system in defining the canonical distribution. It seems to me that the least degree of coupling to well-defined dynamical systems is enough to justify statistical mechanics, not with respect to such gross matters as energy relaxation times, but surely to such subtleties as reversibility. Time's arrow is then the necessary consequence of the fact that no physical theory except perhaps the final one can describe the whole of the universe. It seems also clear that the arrow of time in the sense here described would remain the same for the man who dwells in a contracting, rather than an expanding universe, provided he can once set up, perhaps in some super air-raid shelter, physical systems of the sort we know, temporarily free from large energy inputs out of space. Behind his heaviest shields, gases will leak out of valves irreversibly (unless he pours in the free energy) as they are moved to do by tiny mixing forces out of the external world, however it behaves in the large.

Surely there are other and deeper answers to the problems here touched in an elementary way. But it is worthwhile to try to talk even of these weighty matters in simply physical language, with order of magnitude estimates. There is pleasure and instruction both in such a method. That is what I have learned, however imperfectly, watching with delight the master of the style, V. F. Weisskopf.

REFERENCES

1) J. Willard Gibbs, Elem. Princ. in Stat. Mech. (Dover Press, New York, 1958).
2) N. van Kampen, in: Fundamental Problems in Stat. Mech., (Editor E. Cohen (North-Holland Publishing Co., Amsterdam, 1962).
3) The work of B. J. Alder, of the University of California, Livermore Scientific Laboratory, who has for years been exploring the foundations of the kinetic theory with the computer.

A MATTER OF TIME
P T Landsberg

1. Introduction

I should like to start with a word about my predecessor in this Chair, Professor H.A. Jahn. He was appointed in 1949 as the first Professor of Applied Mathematics in the then University College of Southampton. As Professor, as Head of Department, and as Dean, he has endeared himself to many members of this University. I want to start by wishing him well.

Now a word about solid state theory, in which I have had an interest for 28 years, in fact since I joined Dr. Allibone's team at the AEI Research Laboratory in Aldermaston and also worked with Professor Harry Jones at Imperial College. During twelve happy years at University College, Cardiff, I continued to spend about half my research time on solid state theory. Although I shall say nothing about this important area of work to-day, let me assure you that it is not being neglected by me here in Southampton. In fact, we have already a small but enthusiastic group of young scientists whó (all externally funded, I may say) are looking into problems which arise, for example, in light emitting diodes and in solar cells.

Turning to applied mathematics in general, it yields insights in many fields: the voting behaviour at elections, the orbits of satellites, the shapes of lenses, growth equations in biology, factor analysis in psychology, etc. My main interest is in the applications to physics[1], where with the aid of mathematical models and manipulation (right-hand side of Figure 1) we arrive at relations between concepts. Such relationships have proved exciting by their novelty. A study of light and gravitation has for example led to the discovery of the gravitational bending of light. Other examples are the increase of entropy with time, the impossibility of knowing the position and velocity of a particle with indefinite accuracy, etc. In order to obtain an intuitive grasp of such relations (left-hand side of Figure 1) it is desirable to discard the mathematics temporarily and to think hard about the concepts. With luck, this yields an *intuitive* understanding of what has been achieved mathematically. The discussion of concepts is also outstandingly useful if we want to communicate with a wider audience. Then we discard the right-hand side altogether and develop the left-hand side of the figure. This is always a challenge which calls for new analogies and images. The really good popularisers can do it, our Chairman, if I may say so, being an example.

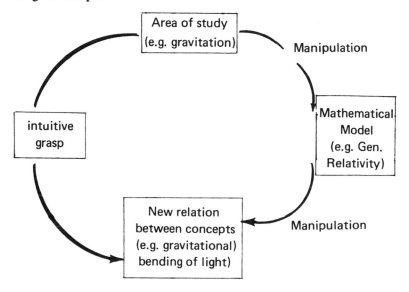

Figure 1.   Intuitive and mathematical grasp of the relation between concepts.

A first excitement comes therefore from the easy manipulation of difficult concepts, and a second from the extrapolation of theories from regions where they are well tested to unknown and distant landscapes, where the tests are difficult or nonexistent.

This type of thing strikes me as the stuff of which inaugural lectures could be made. I have an example here.

2.   Questions of time

A hundred years ago James Clerk Maxwell addressed the British Association in Bradford, saying: "The mind of man has perplexed itself with many hard questions. Is space infinite, and if so in what sense? Is the material world finite in extent, and are all places within that extent equally full of matter? Do atoms exist, or is matter infinitely divisible?

A MATTER OF TIME

The discussion of questions of this kind has been going on ever since men began to reason, and to each of us, as soon as we obtain the use of our faculties, the same old questions arise as fresh as ever"[2]. It is a credit to a hundred years of scientific work that some of these immense questions can now be made more specific, and that provisional answers can be given to others.

Ladies and gentlemen, this is not Bradford and I am, alas, not a Maxwell, but we could perhaps also perplex ourselves with some parallel questions — not about space, but now about time, a topic which has attracted the attention of distinguished British cosmologists[3,4] and of many other scientists, too numerous to name here. How old are the stars (§3), and how old is the universe (§4)? How and when were the chemical elements formed (§5)? Will the universe contract again so that there can be one or many oscillations (§6)? If there are, are they all the same and is life possible in the contracting phase (§10,11)? Is perhaps the progress of time itself an illusion (§§7,8)?

3. Geological and stellar ages

The ages of the earth and the universe are so great that I shall have to scale seconds, minutes, days and years up by factors of one hundred million (10^8) and I shall refer to these vast tracks of time as "long seconds", "long years" etc. (Table I).

We know a great deal about the evolution of stars and clusters of stars. The initial hydrogen is used up at the centre in a kind of nuclear furnace, and the helium already present is further added to as

Table I

Scaled-up units of time

1 long year	100 000 000 years (one hundred million years)
1 long day	274 000 years
1 long hour	11 400 years
1 long minute	190 years
1 long second	3 years

a result of these reactions. For a star of the size of seven suns it takes about ¼ of a long year for the hydrogen to be burnt up. There are theories of the development of the radii and the temperature of stars. From such theories one infers the ages of stars, and these are *at most* of the order of 110 long years[5-7]. Such calculations draw on many branches of applied mathematics, physics and chemistry.

An entirely different method can be used to estimate the age of the earth, and of meteorites. This depends on the radioactive decay of the elements. For example the isotopes U^{235} and U^{238} are expected to have been formed in the earth with equal abundance, so that their different present abundances must be due to the difference in their decay rates. This yields an estimate of about 46 long years for the age of the elements on earth and in meteorites. Some of the moon rocks were brought back to earth for precisely this kind of analysis to be performed on them. They turn out to be of the age of the oldest rocks on earth — about 30 long years old. The oldest pieces of the solar system which we know are certain meteorites (chondrites). They consist of materials that were made 46 long years ago before even the rocks on earth had crystallised.

Let us now turn to a short history of the earth. On our time scale the earth is now a middle aged man, 46 long years old. At age 27 the first forms of unicellular life appeared. At age 44 long years amphibious, land plants and trees developed. It is during the last three or four long years of the earth that deposits of petroleum oil, shale and natural gas began to accumulate. Coal deposits developed only during the last two long years. Mammals appeared during the very last long year, and then only during the last ten long days of the history of the earth. The machine age has been with us for say one hundred ordinary years i.e. for thirty long seconds or so. In this period man has been rapidly using up the reserves of fossil fuels. Another long minute or so and these fossil fuels will be exhausted. This is the real nub of the energy crisis: 2 long years' deposits used up in a few long minutes! But this is another story.

The age of the earth has been a controversial topic. William Thomson (later Lord Kelvin) had studied the Fourier theory of heat conduction, applied it to the cooling of the earth, and had come to the conclusion that at its present rate of cooling the earth could not be more than a fraction of a long year old (1860—89). This was a very inconvenient result. Darwin in his *Origin of Species* (1859) had required three long years to explain certain features of the earth's crust — notably the removal of solid material from chalk cliffs of Kent by water — and for the denudation of the Weald. In geology also much longer times were required by at least some scientists, and seemed inconsistent with Thomson's estimates. This led to a long and famous debate in which great evolutionists like T.H. Huxley took part. Darwin felt forced to modify later editions of his book. Yet Thomson was wrong all the time, not in his calculations, but in his assumptions.

Unknown to everyone a source of energy was at work "in the great store house of creation", heating up the earth and bringing the estimates of physicists into line with those of the biologists and geologists. This source of energy was radioactivity.

Incidentally, in this great geological controversy T.H. Huxley delivered himself of a warning regarding the limitations of mathematics. On the 19th February 1869 as president of the Geological Society of London, he sought to defend the geological time scale against Thomson's theoretical calculations[8]:—

"But I desire to point out that this seems to be one of the many cases in which the admitted accuracy of mathematical processes is allowed to throw a wholly inadmissible appearance of authority over the results obtained by them. Mathematics may be compared to a mill of exquisite workmanship, which grinds you stuff of any degree of fineness; but, nevertheless, what you get out depends on what you put in; and as the grandest mill in the world will not extract wheat-flour from peascods, so pages of formulae will not get a definite result out of loose data." It is a point well taken.

4.   The big bang: expansion ages

It is one of the exciting results of modern cosmology that it arrives at the same sort of time scale as do calculations from the other areas of science, although it is based on an entirely different set of observations. Thus a number of sciences (biology and geology, physics and cosmology) are here supporting each other, showing that knowledge is one and indivisible.

I shall now give you a rough idea of how one can arrive at a cosmological time scale. The set of spectral lines emitted by atoms in incandescent gases are their fingerprints which enables us to recognise and identify different types of atoms, not only on earth but also in the sun and in the light emitted by stars. In the case of galaxies these lines are seen with their correct *relative* frequencies but bodily shifted towards lower frequencies. This effect is well known in the case of sound [the Doppler effect*]. As a train disappears into the distance its whistle seems to have a lower pitch or frequency. In this respect light is just like sound, and one attributes the lowered frequencies of the atoms' radiation to a recession of the emitting source. It was found by the great American astronomer Edwin Hubble[9] that this applies to many galaxies. The universe appeared to be in a state of expansion. But the expansion is not random; he found that the fast moving galaxies are further away than the slower ones. This is consistent with the view that all galaxies including our own were at one time long ago close together. The faster ones shot ahead in the general expansion that has taken place and are now further away from us than the slower ones, which of course lagged behind. The view that the universe is in a state of general expansion has been termed the "big bang" cosmology. It implies that the cosmos is not static, it is not in equilibrium; on the contrary it exhibits systematic motions. This then is the current *belief* — that is until and unless the shift in the spectral lines can be more confidently attributed to another source. Thus the expansion is current orthodoxy, but, like all science, it lacks *certainty*.

We have learnt from history that Rome is not the centre of the universe. Nor is the earth. Nor is the solar system. Hence if *we* see a recession of the galaxies according to Hubble's law, so should observers sitting on *other* galaxies. The law discovered by Hubble ensured just that. It can be visualised by blowing up a balloon on which galaxies are marked by crosses. As it expands, then, whichever galaxy is taken as fixed, *all* others are seen to recede radially from it.

General relativity, using a model of a smeared out universe, in which galaxies are replaced by a uniform distribution of matter, leads to a mathematical description of this process[10]. It is clear that a model of the galactic motion can fix the earlier motions in terms of those observed at present. Hey, presto, an estimate for the time since the big bang must result:

---

*Square brackets surround phrases or topics not mentioned in the lecture.

$$\begin{bmatrix} \text{To-day's} \\ \text{observational} \\ \text{expansion} \\ \text{data} \end{bmatrix} + \begin{bmatrix} \text{Relativistic} \\ \text{(or Newtonian)} \\ \text{theoretical} \\ \text{model} \end{bmatrix} \Rightarrow \quad \begin{matrix} \text{Time since} \\ \text{big bang} \end{matrix} \quad (1)$$

One arrives at about 100 to 200 long years. Though the uncertainty is considerable, this turns out to be greater than the age of the earth and of meteorites — very fortunately — without being an unreasonably great age. There are less standard theories in which Newton's gravitational constant decreases slightly with time. One of my students and I have discussed such things and one finds then that relation (1) leads to decreased ages[11].

Now the *universe* may have existed prior to this big explosion and *its* age is therefore unknown. The above estimates, however, yield the time since the *big bang*, and this is as close as we can get at present to the age of the *universe*.

One can place William Thomson's estimate of the age of the earth as an estimate of the age of the universe in Figure 2[12], on which we also

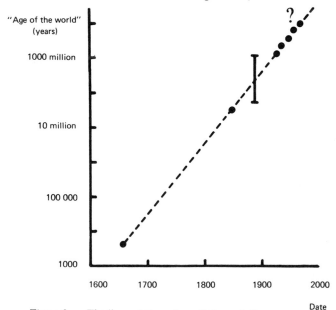

Figure 2.   The "age of the universe" through the ages.

66

show the estimate by Bishop J. Ussher (1581–1656) who claimed to show from the ages of the patriarchs as recorded in the Bible that the universe was created on Sunday 23rd October 4004 B.C. The more recent estimates are based on data from the expansion of the universe. These seem to be lengthening our estimate of the time scale of creation. The vertical line gives a variety of geological estimates for the age of the earth which were current around 1900.

5. The formation of the elements

We cannot trust our science at the very beginning of the big bang: the energy densities are far too high, and this era has to be cut out from our considerations. One might expect that we would have difficulties up to the first hour. In fact, we can do far better and start our investigation within a very small fraction (one ten thousandth) of an ordinary second (10^{-4} sec) (not a *long* second). Things are still pretty hot and dense (Table II) and in fact the energy density is so great that the mass in a cubic millimeter due to radiation alone is 10 million Kg ($\rho \sim 10^{13}$ gm cm^{-3}). This initial hot soup is a mixture of protons, neutrons, electrons,

Table II

The Universe when we can start to calculate

Time $t \sim 10^{-4}$ sec,   temperature 10^{12} °K .
Radiation density ρ is equivalent to 10^{43} nucleons per m^3
Matter density is 10^{34} nucleons per m^3.

positrons, neutrinos and of course radiation. It is so hot that any complete atom is not only stripped of its electrons, but even its nucleus breaks up.

As the universe expands and cools the smaller nuclei form first (Table III) from the ingredients of the initial soup. It is then still too hot for the bigger nuclei to be stable. They form a little later. The whole business of the cosmological element formation (or nucleosynthesis) takes only of the order of one hour and occurs at a temperature

67

of hundreds of millions of degrees.

First hydrogen is formed and then helium. The heavier elements come later, when it is a little cooler, but only in relatively tiny amounts. The helium is produced by (among other mechanisms) the hydrogen — helium reaction typical of the energy-producing reactions in which our

Table III

The first elements to be formed

Element (nucleus)	Isotopes (nucleus)
Hydrogen (1 proton)	deuterium (1 proton, 1 neutron), tritium (1 proton, 2 neutrons)
Helium (2 protons)	He^3 (2 protons, 1 neutron), He^4 (2 protons, 2 neutrons).

researchers at the Culham Laboratory are so interested in their quest for nuclear fusion power. These reactions are responsible for our solar energy, stellar radiation and, as we see now, they are also characteristic of the early universe. Solar energy is free, but fusion energy came first!

This cosmological element production is discussed mathematically by adopting a class of theoretical models for the big bang, and by assuming an initial radiation density, temperature, etc. Then one feeds into the computer the important nuclear reactions which occur at various temperatures, densities, etc. One can then trace the gradual build-up of the elements through more than one hundred and forty reactions. The details are somewhat complicated (Figure 3).

One eventually arrives at theoretical estimates for the amount of the various elements produced in the big bang (Figure 4). These amounts can be calculated down to a radiation temperature of only about 3°K, which is characteristic of the present time, and can then be checked against the known present abundances of the elements, and the present matter density. Present abundances are known from spectroscopic studies, stellar structure analysis and from cosmic rays[5]. In this way

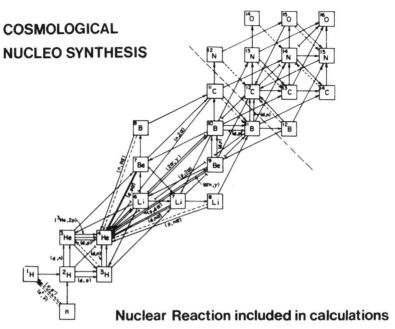

Figure 3. Nuclear reactions included in a recent calculation of element formation in the big bang[13]

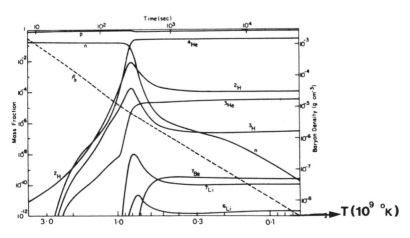

Figure 4. The evolution of nuclear abundances in a typical big-bang model[13]

the *present* matter density is linked through the theoretical model to the amount of helium and deuterium produced in the big bang. These two elements are particularly useful as their incidence is by many cosmologists believed to be largely due to their production in the big bang. There is lots of helium about at present (22%–32% by mass). However for 70 000 helium atoms there exists only one deuterium atom (and only one "helium three" isotope). Even so, one finds that in order that a model of the big bang produce enough deuterium, the present matter density must be small enough[14]. This important and most recently discovered constraint is additional to the need to have the age since the big bang consistent with the known ages of globular cluster stars, and the need for the deduced present mean matter density to be in excess of that known to be in the galaxies alone. In addition the model must be in agreement with the present observed velocity of expansion [which determines the value H_0 of the "Hubble parameter"] and the indirectly inferred slowing-down of the expansion [this yields the "deceleration parameter" q_0]. These two parameter $[H_0, q_0]$ govern the dynamics of the expansion of the universe and are well studied.

When all this has been done, then one is, surprisingly, still left with some very simple models which (for the connoisseur) can be described as matter-dominated, zero pressure, Friedmann type models[15]. The balance at present favours ever expanding models, but under certain conditions (non-zero cosmological constant), oscillating models are not ruled out, particularly if deuterium is produced in supernova explosions, or if matter was withdrawn from the early nucleosynthesis by being bound in black holes.

The relative abundance of the elements in the observable universe is incidentally quite different from that on earth. On earth helium is rare, having largely escaped the weak gravitational field of this tiny planet. In the universe hydrogen and helium together account for almost all the matter.

In all these considerations one achievement stands out: the periodic table of the elements as developed about 1870. Here we are, a tiny speck of matter in a vast universe. Within the last long second we have become familiar with all the chemical elements which can occur either in space or in time, even if we extrapolate back to the initial explosion. Wherever we go, we shall meet the same elements, like familiar, old friends. We shall be able to make water and alcohol and whisky on the rocks. It is a great human achievement to have understood this much, and we can be proud of it (whatever the economic troubles that beset us!) I really think the chemists might make even more of it.

6.   Will the Universe contract?

We must now turn to the long range future evolution of the universe. It is a risky business to expound about the millennia to come when we cannot even be sure of tomorrow's weather.

The question of an eventual contraction versus a continued expansion is surprisingly simple in principle. It is rather like the question of firing a rocket and wondering if it will fall back or go on into space. It will fall back if the earth's attraction is too big for the energy of motion. In a similar way if in our model universe the average density ρ of matter is high enough for an adequate gravitational pull to be exerted, then the universe will again contract. One can follow the numbers involved in this estimate most simply by expressing the density ρ as hydrogen nuclei (or protons) per cubic meter. The present mean density of luminous matter is equivalent to

$$\rho \sim \tfrac{1}{2} \text{ proton per cubic metre}$$

and is too small for a contraction of the universe. But the astronomers have of course difficulty in finding non-luminous matter, so that a better estimate may lead eventually to a density which may be large enough to encourage a belief in an eventual contraction. Professor Pathria and I looked at the condition for contraction last year in the light of the existing uncertainties[16], and found that one would need

$$\rho > 1.8 \text{ protons per cubic metre } [=\rho_c] \tag{2}$$

The time t_0 since the big bang then satisfies

$$57 < t_0 < 163 \qquad \text{(long years)} \quad (3)$$

[We included the possibility of a non-zero cosmological constant and ranges of possible values of the present Hubble constant H_0 and of ρ. This narrows down the usual gap between ρ and ρ_c. This usual gap is obtained by writing $\rho \sim 10^{30}$ gm/cm^3, and taking ρ_c from the Einstein-de Sitter model as $3H^2/8\pi G \sim 10^{-28}$ gm/cm^3. Relation (3) may also be satisfied for an open universe[15].]

71

7.   The arrow of time : the problem

I have talked about the modern cosmology of an expanding universe (§§3—6) and I shall return to this theme shortly (§§10—11). However, I should like to interupt the story at this point in order to explain a few ideas which dominated cosmology at the turn of the century. It will be useful to be familiar with these concepts when I return to to-day's views in the last part of the lecture. Those who are not interested may let their minds wander during this interlude. They should not fail to wake up again when they hear the term *heat death*.

The first question of this flash back is:—

"How is it that there is a direction of time?" This is a funny kind of question you may say, for we all *accept* that there is a direction of time. While that is true, the question is nonetheless justified because there is a complication which I shall now indicate.

Figure 5 shows a billiard ball bouncing off the side of a table (Figure 5(i)). The reversed trajectory is also allowed by the laws of mechanics (Figure 5(ii), and it is precisely the same as the trajectory obtained if time ran backwards (Figure 5(iii)). In fact classical and quantum

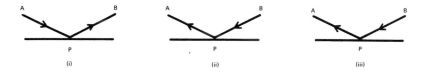

Figure 5.   A billiard ball bounces off a side of a table at P
(i)   It starts at A and ends at B in forward time
(ii)   It starts at B and ends at A in forward time [reversed boundary conditions compared with (i)]
(iii)   As (i) but with time reversed.
Note that (ii) and (iii) are identical.

mechanics teach us that for most elementary process the time-reversed process is possible. Which of these will occur is determined by the so-called *boundary conditions*. In our example the boundary condition is either: "the projection takes place from A" or "the projection takes place from B". A simple illustration of this kind of symmetry occurs when you see your friend's eyes in a mirror, for he can also see your eyes. The laws of reflection are time-symmetrical and the trajectory for rays of light from his eyes to yours is exactly equal and opposite to the trajectory from your eyes to his (as in Figure 5): The laws for the light rays are the same, but the boundary conditions are reversed.

So there is time symmetry for most elementary processes of collisions. Now take a box of gas and isolate it from its surroundings. Take as an initial (or boundary) condition that the ends are at different temperatures. What will happen? "Simple", you say, "the gas will attain a uniform temperature." That is correct; but remember *how* this happens. The molecules collide and exchange energy, and the faster molecules from the hotter parts will tend to lose energy to the slower molecules from the cooler part. Each collision is time symmetrical so that the reversed process is also a possible one. We have:

$$\text{temperature difference} \rightarrow \text{collisions} \rightarrow \text{uniform temperature}, \qquad (4)$$

One might also expect the reverse process:

$$\text{uniform temperature} \rightarrow \text{collisions} \rightarrow \text{temperature difference} \qquad (5)$$

If (5) occurred as easily as (4) we would have time symmetry in complex processes as expected from the elementary processes. In fact, (4) is overwhelmingly more frequent than (5). Our question "Why is there a preferred direction of time?" is another way of saying: "Since there are mainly time-symmetrical elementary laws, how can they give rise to the non-time-symmetrical processes (4) of heat conduction, and diffusion, in systems which are left to themselves?" It is a basic question, it has been around for a long time, but from the shelves of books and papers which have been written on it, no concensus of opinion has emerged.[17-20]

One idea is to note that there is a sense in which all that happens can be traced back to the beginning (or end) of the universe, since the whole of evolution, human history and literature is a part of the basic cosmological process. In this rather obvious sense time asymmetry itself is also dependent on cosmology and the cosmological boundary conditions. However, this hook-up between cosmology and heat conduction seems somewhat artificial.

[There is one proviso which should be attached to this formulation of the problem. In certain elementary particle reactions (neutral kaon decays) the time reverse process never happens and one should check that this does not alter the problem posed. One can show that in these systems the generation of spontaneous temperature differences by process (5) is again a possible process, but now only in the corresponding system of *anti*particles. We are therefore still left with the problem of explaining the non-occurrence of processes (5) as against the occurrence of processes (4)[21]. General relativity, too, furnishes (in the Friedmann models) universes which are time symmetrical. A preferred time direction is singled out by empirical data (Hubble expansion) which suggest therefore the need to invoke a boundary condition.]

8.   The arrow of time : a solution

In answer to our question regarding the arrow of time consider Figures 6 to 8. A partition confines a gas to one half the volume of a container (Figure 6). Immediately after withdrawal of the partition, the gas (if it is not interfered with) is in the state of Figure 7, and drifts to a state such as that shown in Figure 8. This drifting implies what we call a non-equilibrium situation. Now everybody knows that Figure 7 will lead to Figure 8, whereas the reverse is never seen (if there are enough molecules in the gas). But to the purist what is never seen could still be possible, though we may have to wait a thousand long years for it; for the reverse process is compatible with the laws of mechanics because of the time reversibility of collisions. We can make a verbal allowance for this by calling it a *fluctuation from the equilibrium state*. Our second question in this interlude is: What principle enables us to call the process from Figure 8 to Figure 7 a *fluctuation*, and the reverse process a *normal* one?

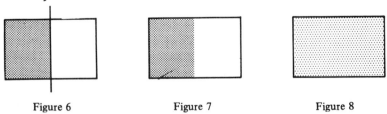

Figure 6   Figure 7   Figure 8

A system in an equilibrium state (Fig. 6), after the withdrawal of a partition (Fig. 7), and then left to itself, will go over into another equilibrium state (Fig. 8)

In order to answer this, divide the box into little volumes of a cubic centimetre each. Let us generate Figure 6 by starting with all these little volumes empty and placing the available molecules one by one into the little volumes in the left-hand box until we have the right number of molecules, more or less uniformly distributed. Some little cubes will then be empty to allow for the space between the molecules, and each arrangement of empty and full boxes is a *realisation* of the state of Figure 6. But the number of such realisations is usually vast, and we have no experiments to distinguish between them. So we lump them altogether by simply counting them, and we then give this number a name. Let us call it the *realisability* of the state. Let us go through the same process for the state of Figure 8. There are now many more little cubes available to be filled, so that the number of ways of realising this state is enormously greater for Figure 8 than it is for Figure 6: The *realisability* of the state of Figure 8 is enormously greater than that of Figure 6.

The basic principle at work here is this: Suppose a system develops without interference from the outside. Then it chooses among its available equilibrium states in proportion to their realisabilities. I shall call this principle **P** [22], and we note that it makes at least no explicit reference to time. But it implies that if one available state is enormously more realisable than any other, then we do not go far wrong by predicting that this state will be *the* equilibrium state. Thus if the partition is removed from a state such as that of Figure 6, the principle tells us that the system will go to a state such as that of Figure 8 *because that state has the greatest realisability*. In this way, then, the principle extracts a direction of time even though the molecular collisions which give rise the diffusion of the gas are each time-reversible. It does so by statistical averaging in proportion to the realisabilities.

After writing this section I dreamt that I encountered a superhuman intelligence in the guise of some shadowy demon. He turned on me and thundered: "You, little man, average because you are ignorant. I have a thousand eyes and a million brains and I can take in at once which of your stupid little cubes is empty and which is occupied. It takes no effort at all on my part to distinguish any two arrangements of molecules over volumes. Lumping this vast number of realisations of the state of Figure 8 together is to remove all the individuality of these arrangements which *I* can see, even if *you* cannot. You", he added scathingly, "have to do this to create a crutch for your weak little brain — but you have achieved nothing." I was downcast in my dream, but replied it as dignified a way as possible: "To you, Sir, the realisability of each state is unity, for you know all, and you distinguish all. So the principle **P** which extracts a direction of time for me, is not needed by you." After some reflection I pressed on: "In fact, to anyone with all *that* knowledge, the universe must be stretched out in front of him to be comprehended with all its history and in all its detail in a single leap of your powerful mind. Perhaps the principle **P** is meaningless for you, because you do not require the notion of a direction of time?" It was a question. But the demon did not reply; he had disappeared (rather conventionally) in a puff of smoke.

[The fine-grained entropy $\mathrm{Tr}(\rho \ln \rho)$ is independent of time.]

This demon lives in a world of molecular swarms, of atomic collisions, of seas of electrons. He knows nothing of human concepts such as table, cloud, books — he regards all these as rough, inexact terms for certain groups of molecules required by the rough and inexact sense data of humans. The demon would have to be taught not only our language, but our very concepts. For him, therefore, a direction of time would emerge only as a result of averaging over many of the states which he can in fact distinguish. He has no *need* to average, whereas we average over states we cannot distinguish. It makes one think that at this level the direction of time is a large-scale (macroscopic) concept and basically a human illusion of the same type as the belief, suggested by our raw sense data, that this table is smooth, solid and without structure. To sum up: In the language of ordinary experience time has a direction in virtue of principle **P**, but, for more penetrating observers, the direction of time has to be *derived* by averaging. [However, if one mixes the languages of the human observer and of the super-human observer paradoxes of various kinds can arise[23].]

For our purposes the idea of realisability of a state is called the *entropy*. As an isolated system progresses in time we see that its entropy increases. Some will recognise this to be part of the second law of thermodynamics which, as cultured people, I understand we should all know about.

[But there are dangers. One can realise this if one recalls that the great American historian Henry Adams regarded human thought and human history as similar to a liquid or a gas, and as passing through phases like ice, water and steam during the processes of history. Though I do not exactly support his view, Henry Adams has endeared himself to me by saying that "the future historian must seek his education in the world of mathematical physics"[24].]

9.   The doctrine of the heat death

This last interpolation about older cosmological thought leads us now to the idea of the *heat death* of the universe which was current at the end of the last century. Thermodynamics was one of the great excitements of that period. Thermal equilibrium and maximum realisability, or entropy, had been discovered as the end point in the development of any isolated system. The universe was regarded as no exception. It was thought of as a static system which would eventually consist of dead bodies gravitating in a space devoid of life. This is the doctrine of the heat death.

[Lord Kelvin in his paper on the dissipation of mechanical energy into heat[25] also pointed out that the earth has been, and will again be, unfit for human habitation subject to nature being roughly constituted as then known. The idea of a heat death was made explicit by Hermann von Helmholtz in a lecture on the interaction of natural forces given in Königsberg in 1854. The idea remained around for a long time: ". . . the thermal form of energy used often to be called dissipated or degraded energy, so that the second law proclaims a steady degradation of energy until all tensions that might still perform work and all visible motions in the universe would have to cease. All attempts at saving the universe from this thermal death have been unsuccessful, and to avoid raising hopes I cannot fulfil, let me say at once that I too shall here refrain from making such attempts."[26]]

This depressing prospect concerning the future painted by the physicists of that era contrasted sharply with the optimism generated by the new belief in progress by Darwinian evolution. But this pessimism was occasionally enlivened by some diverting thoughts. For example, the suggestion was made in 1852 that energy might be reflected at the boundaries of space in order to recreate temperature differences[27] — a suggestion not rendered more realistic by the fact that it was made by a civil engineer. Maxwell, in 1867 introduced a demon to achieve the same effect. This demon was sitting in a gas and, by means of a partition with a shutter, he could let the faster molecules enter one half of the container, and the slower ones the other half. Now a system with faster molecules has the higher temperature. Maxwell had thus made plausible the generation of temperature differences spontaneously and systematically — though we know very well that this never occurs. In fact it was shown later that the radiation energy needed by the demon to recognise the molecules made this set-up useless for the production of mechanical work. But Maxwell's main object in creating him was to show the statistical nature of the law of entropy increase, as he pointed out in his "catechism on demons"[28].

These discoveries, by means of Maxwell's demon and by other less exotic methods, furnished the basis for the views concerning time's arrow which I have already explained: there is the usual dominant direction for diffusion and for temperature equalisation, but it is tempered by the possibility of fluctuations. The possibility of fluctuations arose from the statistical interpretation of entropy increase, and it is these fluctuations which provided the hope of an escape from the heat death. Indeed, if after the heat death our disembodied intelligence were to wait for an almost interminable track of time, a giant fluctuation could in a static universe bring into existence another solar system and another earth, possibly as an exact repetition. Shelley has a relevant passage:

> The world's great age begins anew,
> The golden years return,
> The earth doth like a snake renew
> her winter weeds outworn.
>
> A loftier Argo cleaves the main,
> Fraught with a later prize;
> Another Orpheus sings again,
> And loves, and weeps, and dies;
> A new Ulysses leaves once more
> Calypso for his native shore.
>
> Another Athens shall arise,
> and to remoter time
> Bequeath, like sunset to the skies,
> the splendour of its prime;

Loschmidt's reliance on fluctuations carried him, however, too far:

"Mankind," he remarked, "would have an inexhaustible supply of transformable heat at hand in all ages." What he did not realise is that the fluctuations cannot be used for the systematic production of mechanical work.

I will not dwell here on all the problems (and the considerable literature) which the idea of a heat death created for the theologians. It was clearly, difficult to see how human values could have a high status in a cosmological process which led to their destruction, or how God could have any meaning in a world devoid of life. But Christian apologetics did better than that. Some argued that since the universe was not yet in thermal equilibrium it must have existed for only a finite time; and this led to the idea of a creation and hence it led back to God after all[29]. Suffice it to reiterate here that the existence of life is compatible with a static universe which is old enough to have attained overall thermal equilibrium, by supposing that we are currently involved in a fluctuation from that equilibrium[30].

All these nineteenth century cosmological ideas assumed a static universe, contrary to what we believe now (§4).

10. Speculations about an oscillating universe

Returning to modern cosmology, we have already seen (§6) that the expansion may continue indefinitely. This is the modern form of the doctrine of the heat death, since it would lead to a cold, dead universe. However, it is just possible that the present expansion is part of an oscillating universe. The recurrence of an earlier part of the history of the universe, in *some* form yet to be discussed, is the modern counterpart to the recurrence by a giant fluctuation in a static universe, which I noted when I cited Shelley.

Now, assuming oscillations, are they all identical? In fact is the oscillating universe nature's largest clock? I believe this is not so. Some detailed calculations by Tolman[31] and by Professor David Park and myself[32,33] suggest that the cycles of an oscillating universe become longer and longer, as if the universe had more and more difficulty in contracting again. This means also that the more cycles have gone by, the more difficult is it to recognise the universe to be oscillating.

Let us now ask: "Why is it difficult for current cosmology to decide if the universe is ever-expanding or oscillating?" Our answer is: "Because the universe is old, and though oscillating, it has already gone through many cycles."

[This reason has only recently been proposed in the light of detailed calculations[33].

If there is a recontraction to another big bang one could ask: "How long will it take?" The answer to this question is given by the simple formula (ρ_O = present matter density, ρ_C = minimum density for contraction to be possible)

$$T = \frac{\pi}{H_O} \frac{(\rho_C/\rho_O)^{1/2}}{(1-\rho_C/\rho_O)^{3/2}}$$

which is difficult to evaluate because of the uncertainty concerning ρ_O. Perhaps we are one tenth through the cycle from big bang to big bang, but this is only an inspired guess, and who knows how many big bangs there have been before?

The regimes of extreme compression occur when the end of one cycle merges into the beginning of the next. I would argue that one must leave a blank here and not attempt too detailed a physical picture of this regime. For it is quite possible that our science is not able to guide us in conditions which are utterly different even from those believed to obtain in the hottest stars.]

11. Life in a contracting phase

I shall now become even more extravagent and abandoned to distant dreams, for I shall consider briefly life in a contracting phase. It is of course possible that there is no contracting phase but indefinite expansion, or that there would be no life if there were a contracting phase. This is a *first* possibility. However, a more interesting *second* possibility is that fairly normal life is possible in a contracting phase.

We have already noted (§4) that the light from receding galaxies is shifted to lower energies and this is the situation for an expanding universe. The decreased radiation energy on earth in an expanding universe implies the darkness of the night sky. Space is for us a great absorber of radiation — an important and interesting point which has been emphasized in recent years particularly by Sir Hermann Bondi, whose slim and elegant book on cosmology was a pioneering effort of the 1950's. That was a time when it was still possible to write books on this subject which were less than 500 pages long.[34] For a contracting universe the reverse is true. The light from the distant galaxies is jacked up in energy and focussed by a kind of head light effect and the night sky would be bright. This fact alone may have important consequences for the evolution of life. Nonetheless, while this is not "life as usual", life in some form would seem possible.

A *third* possibility was implied by Boltzmann[30]. He supposes life in the present epoch to be a fluctuation, as already observed (§9). Living beings, he remarks, will regard the direction of time towards smaller entropy states as "the past", and that towards higher entropy states as "the future." On this basis, if the entropy, in the contracting phase returns to the value it had at the beginning of the cycle, we would then have the situation shown in Figure 9. In the expanding phase time runs forward for us, and it reverses in the contracting phase, with the amusing consequence that human beings would still regard the universe as expanding. Only our disembodied intelligence could know what was happening. Such an intelligence could not be confined in a box of gas like Maxwell's demon! On the contrary, it would be the universe's great outsider. Anyway the arrow of time, according to these ideas, reverses for us as the universe collapses again[17].

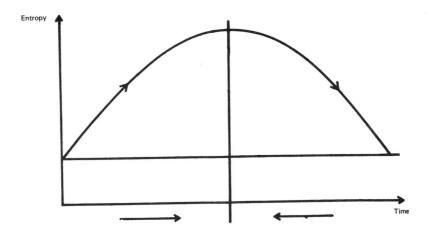

Figure 9. A contracting phase follows expansion. The arrows on the curve correspond to the time as it appears to a disembodied intelligence. The arrows below the curve indicate a conjectured direction of human time.

This amusing (*third*) possibility hinges on the idea of a drop in the entropy in the contracting phase. However, I prefer to suppose that the entropy increases all the time throughout a cycle even in the contracting phase, as already discussed (§10), since this is suggested by model calculations in which one can actually define an entropy[33]. If this is accepted, we would have to rule out this third possibility and we would be left with the view that if there is a contracting phase, and life is possible in it, then it would be more or less the sort of life we know, and the vision of exact repetitions in each cycle would be unlikely to be realised.

[These considerations lead one to wonder if the class of living processes cannot be defined in some way by reference to time reversal[22] — but this is another story.]

12. Conclusion

I have come to the end of my questions of time. We have seen that there are some outlines of answers to these questions.

There are of course other puzzling and exciting matters of time which it is intriguing to contemplate, but for which I have neither adequate time nor adequate insight. If, as I have suggested, a direction of time emerges for larger systems, does it jump about in some sense for very small systems? Or are there some indivisible bits of time as there are atoms of matter? Another problem is that the natural world suggests to us via relativity that space and time should be treated on the *same* footing, while our *consciousness* makes a clear distinction between them. Again, the notions of *purpose* and *plan* do not seem to have any simple explanations within physics. Perhaps this radical difference between biology and the natural sciences can tell us something about the nature of life, as I have already suggested (§11).

We should also ask if we can even imagine situations in which the concept of time itself may not be needed. It is fairly widely known that if a space traveller moves at a high speed relative to the earth, he ages less than his earth-bound friends. Let us add a fanciful extrapolation, namely that the traveller can approach the speed of light very closely. We would then regard his clocks as having been arrested almost completely, and he could visit us again and again almost untouched by the ravages of time. To such a traveller everything would happen almost at once. His clock would go more and more slowly as he approached

82

the velocity of light. Indeed a free packet of light energy, a photon, could by reflection between mirrors visit the same place again and again in the same instant of its own time. If one wants to make a model of our disembodied and all-knowing intelligence, he could be thought of as just such a fast traveller. I should like to leave you with the thought that he might not need the notion of time. Here, in the virtual disappearance of time, Faustus would find hope at last:

> Stand still, you ever-moving spheres of heaven,
> That time may cease, and midnight never come;
> Fair Nature's eye, rise, rise again, and make
> Perpetual day; or let this hour be but
> A year, a month, a week, a natural day,
> That Faustus may repent and save his soul!
> O lente, lente currite, noctis equi!

A MATTER OF TIME

REFERENCES

[1] P.T. Landsberg, *Entropy and the Unity of Knowledge* (Cardiff: Univ. of Wales Press) 1961 and 1964.

[2] J.C. Maxwell, Molecules. Nature 8 437 (1873).

[3] H. Bondi, *Physics and Cosmology,* Observatory 82 133 (1962).

[4] F. Hoyle, *The Asymmetry of Time,* Third Annual Lecture to the Research Students' Association, Canberra 1962 (Canberra: Australia National University) 1965.
W.H. McCrea, Review of[20] in Nature **253,** 485 (1975).
G.J. Whitrow, *The Natural Philosophy of Time* (London : Nelson) 1961.

[5] Y.B. Zeldovich and I.D. Novikov, *Stars and Relativity* (Chicago: Univ. Press) 1967; P.J.E. Peebles, *Physical Cosmology* (Princeton: Univ. Press) 1971.

[6] S. Weinberg *Gravitation and Cosmology.* (New York : Wiley) 1972, p.188.

[7] D.W. Sciama, *Modern Cosmology* (Cambridge : Univ. Press) 1971.

[8] S.P. Thompson, *Life of Lord Kelvin* (London: McMillan) 1910, Vol.I.. p.546.

[9] E.P. Hubble, A relation between distance and radial velocity among extragalactic nebulae, Proc. Nat. Acad. Sci. **15,** 169 (1929).

[10] A. Einstein, Kosmologische Betrachtungen zur allgemeinen Relativitätstheorie, Sitzungsber, preuss. Akad.d.Wiss. Berlin, 142 (1917). Reprinted in English in
    (a) H.A. Lorentz, A. Einstein, H. Minkowski and H. Weyl, *The Principle of Relativity* (London : Methuen) 1923
and in
    (b) C.W. Kilmister (Ed.), *The General Theory of Relativity* (Oxford: Pergamon) 1974
A. Friedmann, Uber die Krümmung des Raumes Z.f. Phys. **10** 377 (1922).

[11] P.T. Landsberg and N.T. Bishop, A principle of impotence allowing for Newtonian cosmologies with a time-dependent gravitational constant, Monthly Notices of the Roy.Astr.Soc. **171,** 279 (1975).

[12] Based on G. de Vaucouleurs, The case for a hierarchical cosmology, Science **167.** 1203 (1970).

[13] R.V. Wagoner, Big-bang nucleosynthesis revisited, Astrophysical Journal **179,** 343 (1973).

[14] J.M. Pasachoff and W.A. Fowler, Deuterium in the universe, Scientific American May 1974;
D.N. Schramm and R.V. Wagoner, What can deuterium tell us? Physics Today, December 1974.

[15] J.R. Gott III, J.E. Gunn, D.N. Schramm and B.M. Tinsley, An unbound universe? Astroph. J. **194,** 543 (1974).

[16] P.T. Landsberg and R.K. Pathria, Cosmological parameters for a restricted class of closed big-bang universes, Astrophys. J. **192,** 577 (1974).

[17] P.T. Landsberg (Ed.), *International Conference on Thermodynamics, Cardiff, 1970* (London: Butterworths) 1970.

A MATTER OF TIME

[18] J.T. Fraser, F.C. Haber and G.H. Müller (Ed), *The Study of Time* (Berlin: Springer) 1972.

[19] B.Gal-Or (Ed)., *Modern Developments in Thermodynamics* (New York : Wiley) 1974.

[20] P.C.W. Davies, *The Physics of Time Asymmetry* (London: Surrey University Press) 1974.

[21] Y. Ne'eman, Time reversal asymmetry at the fundamental level-and its reflection on the problem of the arrow of time; A. Aharony Microscopic irreversibility, unitarity and the H-theorem. Both papers appear in reference 19.

[22] P.T. Landsberg, Time in statistical physics and relativity, Studium Generale **23**, 1108 (1970); also in ref. 18.

[23] P.T. Landsberg, *Thermodynamics with Quantum Statistical Illustrations* (New York: Interscience) 1961.

[24] H. Adams, *The Degradation of Democratic Dogma.* (New York: Macmillan) 1919 and 1947.

[25] William Thomson, On a universal tendency in nature to the dissipation of mechanical energy. Phil. Mag.**4**, 304 (1852).

[26] L. Boltzmann, Address to a formal meeting of the Imperial Academy of Science, 29th May 1886. Translation in L. Boltzmann *Theoretical Physics and Philosophical Problems.* (Ed. B. McGuiness) (Dordrecht: D. Reidel) 1974, p.19.

[27] J.M. Rankine, On the reconcentration of mechanical energy of the universe, Phil. Mag.**4** 358 (1852).

[28] C.G. Knott, Life and Scientific work of Peter Guthrie Tait (Cambridge: University Press. 1911) quotes on p.213 a letter from J.C. Maxwell to P.G. Tait, dated 11th December 1867 which makes the first mention of Maxwell's "demon". See also J.C. Maxwell *Theory of Heat* (London: Longman's Green & Co.) 1871, p.308. J. Loschmidt, Der zweite Satz der mechanischen Wärmetheorie Akad.d.Wiss. Wien, Math.phys.Kl.**59** (Abt.2) 395–418 (1869) arrived at related conclusions. See also E.E. Daub, Maxwell's demon, Hist. Phil. Science 1 213–217 (1970).

[29] J. Schnippenkötter, *Der Entropologische Gottesbeweis* (Bonn: A. Marcus and E. Webers) 1920.

[30] L. Boltzmann, *Lectures on Gas Theory,* (Berkley: Univ. of California Press) 1964, p.447. Translated by S.G. Brush. The original volume II was published in 1898.

[31] R.C. Tolman, *Relativity, Thermodynamics and Cosmology* (Oxford: Univ. Press) 1934.

[32] D. Park, Entropy in an oscillating universe, Collective Phenomena **1**, 71 and 111 (1973).

[33] P.T. Landsberg and D. Park. Entropy in an oscillating universe (Proc.Roy. Soc. A (1975) to be published).

[34] H. Bondi, *Cosmology* (Cambridge: University Press, 1952).

Part B

Cosmology and Electrodynamics

V. *The end of the universe*. The Bathos, by William Hogarth (1764). Dedicated to the Dealers in Dark Pictures, Hogarth here ridicules absurdities sometimes seen in serious pictures and old masters. He also brings together many objects denoting the end of time. The figure of Time leans in agony against the wall of a crumbling tower. His will in one hand leaves all to Chaos, his sole executor. There is a worn out scrubbing brush, a cracked bell, a fractured palette, a man hanging in chains, a darkened room, the broken chariot wheels of the sun, an empty purse, a decaying cottage and a withered tree. The plate was intended as a tail-piece, in contrast to a frontispiece, to a collection of his works. Hogarth died in October 1764 and he regarded his own decline as equivalent to the end of the world—the death of Time.

90

FRED HOYLE, F.R.S.

THE ASYMMETRY OF TIME

PART I

1. *Universal time as a grid number*

A particular point on an Ordinance Survey map is described by two
numbers or coordinates, the well-known grid numbers. Because two
numbers are needed for each point, the map is said to be two-
dimensional. The whole Universe can be thought of as a four-
dimensional map because four numbers are needed to specify each
point.

A walker consulting an Ordinance map is in no doubt that all the
places on his map exist; he does not think of existence as being confined to the particular place where he happens to be standing. Similarly,
the scientist regards all points of the four-dimensional Universe as
existing everywhere with equal reality. Since one of the four numbers
belonging to the four-dimensional map represents 'time' (the other
three represent 'space') we see that all moments of time exist with
equal reality. The future and the past are every bit as much a part of
the map as is the present. It is we, in our capacity as observers, who
single out the present, just as the walker singles out the particular
point where he happens to be standing.

The analogy of the Survey map can usefully be followed further.
We all know when we consult a map that events not marked on the
map are taking place — we can, for example, point to the position of
some pub, but a knowledge of the appropriate grid numbers will not
tell us what is happening inside the pub. The grid numbers simply
describe the locations of events. At one place a tree might be waving
in the wind, at another a man might be downing a tankard of ale. In
the same way the four coordinates of the Universal map serve to
locate physical quantities, a particle, an electric field. . . . These can
be spoken of as elementary events: each has a location specified by a
set of four numbers. At each point of our Universal map we have a
collection of elementary physical events, just as on our Ordinance map
we have perhaps grass at one point, a house at another . . . .

The physical events associated with different location points of the
four-dimensional map are not all independent of each other. If every

point could have events independent of those at all other points, the world would indeed be 'full of sound and fury, signifying nothing'. It is just the interconnection of events, one location with another, that gives regularity and order to the world. The study of these interconnections is the whole business of science. When we discover some rule of interconnection (or think we do) the rule is referred to as 'a law of physics'. The laws of physics represent *restrictions* on the way in which the pattern of physical events can be laid down on the four-dimensional map.

Let me give one or two examples. Suppose at a particular location there is an elementary event corresponding to a particle of particular sort, say an electron. Then in the neighbourhood of that location there must be other locations where an electron exists. If we take all such locations, they form a line. That is to say, a particle cannot be confined to a point, it forms a line in our map. Similarly an electric field cannot be confined to a point. If an electric field exists at a point, then an electric field must also exist in the neighbourhood of that point. In this case, however, the locations form a cone in the four-dimensional map, not a single line. Furthermore, if at a point we have both an electron and an electric field, the corresponding line and cone are not independent of each other; they are connected in the mathematically precise way imposed by the laws of physics. (Language is imperfect here. It is not of course the physicist's laws that produce the interdependence. The Universe itself exhibits the interdependence. The physicist has simply inferred his law from what he finds.)

The last way in which I wish to use the Ordinance map analogy is this: The particular grid numbers that one normally uses have no special significance. It would be possible to swing the grid round through an angle. As long as one specifies the angle it is easily possible to pass from the new grid numbers back to the old numbers. The point I wish to make is that after swinging the grid the same place on the map will in general be referred to by a different pair of numbers. The territory remains the same but the numbers change. Or one can change the map projection. The use of the same grid on different projections (Mercator, Polar, Zenith, etc.) associates different numbers with the same place. Obviously this has no effect on the territory itself. Even if one tries to be a bit more 'absolute', by considering the oceans and continents on the surface of a globe, the normal specification of longitude depends upon drawing the zero meridian through Greenwich. Any other place could have been used. And latitudes would be changed if instead of using the north and south poles we elected to use some other pair of diametrically opposite points. In all these cases, so long as we specify how we are changing to our new system, it is possible to relate the new coordinate numbers to the old numbers by mathematical relations, mapping relations as they are often called.

An exactly similar situation applies to the four-dimensional map of the Universe. It is possible to describe locations in many different ways. The location of the same physical events can be described by different sets of numbers in different systems. Why fuss about this? Because one of the four numbers represents 'time' and the other three represent 'space'. We see immediately that the Universe has no absolute time system – no 'ever rolling stream'. An infinity of universal grid systems are possible, each with a different 'time'. Just as two points on the Earth with the same normal latitudes will not have the same latitudes if one changes from the normal poles, so two different locations may have the same 'time' in one grid but not the same time in another. There is no absolute simultaneity.

Another crucial point appears if we go back to the restrictions that make an ordered Universe possible. These restrictions must be the same for all grid systems, just as the actual continents and oceans are quite independent of what system of latitude and longitude we set up. *The restrictions are absolute.* This is Einstein's Principle of Relativity (badly named I think). It had always been recognized that the special choice of the three 'space' numbers can have no effect on the laws, and this was already incorporated in Newton's work. Einstein's contribution lay in extending what had been done for the three 'space' numbers to the whole set of four numbers, space and time together. Before Einstein, it was as if one had admitted the arbitrary character of longitude (the Greenwich meridian) but not the arbitrariness of latitude (the possibility of changing the poles). Newton's laws represent restrictions that are independent of changes in the spatial part of the grid, but they are not independent of changes in the time part. Einstein insisted that the laws be altered to a form independent of changes in the time part. His modifications implied different restrictions, restrictions which have since amply been confirmed by experiment. Einstein's great success makes it clear that universal 'time' is a mere grid number not an absolute quantity.

2. *Personal time as an absolute or invariant quantity*

It is obvious that if one walks across the countryside along some assigned track the distance from a particular starting point to a particular end point is quite independent of the grid. Grid numbers can in fact be used to calculate such a distance, but the answer always turns out the same, whatever the grid. The distance is said to be an invariant. The same is true of the universal map. Distance along any track is always the same, irrespective of the grid.

Although a human being is really a complex collection of particles, it is possible to think of ourselves, for ordinary macroscopic purposes, as single particles. Our existence then consists of a line in the map, not in general a straight line, by the way (the concept of 'straight' can readily be formulated in mathematical terms). Distance along our line can be determined, and this distance measures our personal or 'proper' time. A 'clock' is a cyclical device that measures equal distances along our line when we carry it with us. Once we know how much distance corresponds to each cycle of the clock, the distance from one point of our line to another is obtained simply by counting the number of cycles performed by the clock in the relevant section of the line. We can think of ourselves as making a journey across the universal map. As we make the journey the clock goes repeatedly through its cycles; a simple counting tells us how far we have gone. One makes such a journey even if one sits in an armchair. A man sitting all day, another running all day, have journeys across the four-dimensional map that are very little different from each other. Age is determined by the length of one's track: the longer the track the older one is. Metabolic processes follow this personal time, because such processes have the properties of coarse-grained clocks.

Physicists idealize the actual state of affairs. An 'observer' is taken to be deathless and birthless so that his line has no end and no beginning – unless the universe itself had a beginning, in which case the line ceases when it comes to the 'edge' of the map. Clocks are also idealized. The observer's clock measures meticulously equal distances along his line. It never needs oiling or repairing! Practical clocks do in fact approximate pretty well to an idealized clock. A good wristwatch ticks off equal distances to an accuracy of about one part in a hundred thousand, while the best practical clock does so to about one part in ten thousand million.

As an aside, we might notice that it is possible to choose the grid system for the universal map in such a way that universal grid time along the line belonging to a particular observer agrees with that observer's personal time. The observer may then make the mistake of thinking that because his personal time has absolute significance so has the grid time. This indeed is the mistake of Newtonian physics. The mistake arises because no grid agrees with the personal time of every possible observer. Except in special cases two different observers require quite different choices for the grid.

94

In the first section I explained the importance of restrictions. What are the restrictions on the possible lines in the map that an observer can have? To answer this critical question I must now bring out an important difference between the universal map and an ordinary map. On an ordinary map there is always a positive distance between any two neighbouring points. On the universal map it is possible for two neighbouring points to be separated either by a positive or a negative

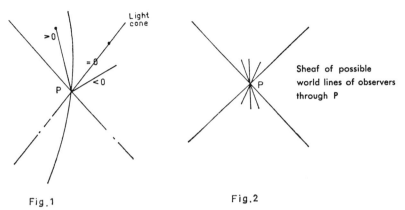

Fig.1   Fig.2

distance. Let P in Fig. 1 be any point of the map. What decides whether the distance to a neighbouring point shall be positive or negative? Suppose light is emitted by a source at P. The source creates an electric field which radiates along the cone shown in Fig. 1 – this is the cone mentioned in Section 1. The cone has the property of separating positive from negative distances. If Q lies inside the cone the distance from P to Q is positive, if Q lies outside the cone the distance is negative – if it lies along the cone the distance is zero! Distances along any observer's line must always be positive. If P lies on the line of an observer then the *whole line must lie inside the cone at P*. P can be any point on the observer's line.

One can think of a number of observers coming together – i.e. of their lines intersecting each other. Suppose this happens at P of Fig. 2. What are the possible lines for observers at P? The answer is again provided by the cone of light emitted from P. Any step from P that lies within the cone represents a displacement along the line of a possible observer. What decides which member of the possible sheaf of lines belongs to a particular observer? The state of motion of the particle when it is at P. The fact that only the sheaf of lines within the cone is permitted is sometimes expressed by the rather loose statement that no particle can move faster than light.

The states of motion of a particle at P decide the form of the line in the neighbourhood of P, but do not decide how the line continues onwards into the cone. This depends on the forces to which the particle is subsequently subjected. If an observer possesses a powerful rocket attachment, then by firing the rocket appropriately it is possible to exercise a considerable measure of control over the way the line continues from P. But no rocket can enable the observer to break the restriction discussed above.

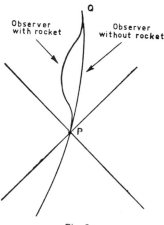

Fig. 3

Suppose two observers are together at P, as in Fig. 3. Suppose one of them to be equipped with a rocket which he uses in a way that causes the two lines first to diverge, then to come together again at Q. Will the two observers have aged by the same amount when they meet? The answer depends on the lengths of the two lines going from P to Q by their different routes, and this is determined simply by counting the cycles of similar clocks carried by the two observers. A glance at Fig. 3 suggests it to be unlikely that the lengths will be the same, and this turns out to be correct. The observers in fact age by different amounts. To decide which observer will be younger one must calculate the distances involved, using the relevant mathematical apparatus. This leads to a result that Fig. 3 does not prepare one for – the observer with the rocket has the shorter track. He is the younger when the two meet. Figure 3 suggests the opposite, but Fig. 3 is drawn only for general illustrative purposes; it does not show negative or zero distances. Remembering that steps along the cone have zero distance, one realizes that the observer with the rocket really does go a shorter way, *because his track takes him nearer to the cone.*

This is precisely the case of the space traveller who leaves the Earth, travels into space, then reverses his motors, and returns to Earth. The traveller is the younger. Calculation shows that if his rocket subjects him to an acceleration equal to twice normal gravity throughout the whole of his trip (except during the brief moments of reversal), and if he ages by forty years during his flight – his clock ticks off forty years – then on his return he will find the Earth to have aged by 12,000,000 years! This effect is sometimes referred to as 'the clock paradox', although it is no more a paradox than is the fact that different routes from one point to another on a Survey map require one to travel different distances. Going from London to Manchester via San Francisco is quite a different matter than travelling by train from Euston. The notion of paradox really springs from the mistaken Newtonian idea of an absolute universal time grid. Both observers start with the same grid numbers at P, and they also have the same grid numbers at Q. But the grid numbers are only a means of describing the locations of P and Q. It is true that the grid could be specially chosen to coincide with the personal time of either one of the observers. But it will not then coincide with the personal time of the other, and it is the latter that determines the ageing.

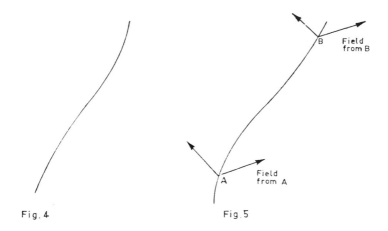

Fig. 4   Fig. 5

3. *Forces and communication between material systems*

The line of a particle in the map is influenced by forces. More precisely by fields, of which the electric field of Section 1 is one example. In Fig. 4 we have the line of a particle. What counts so far as this particle is concerned is the field events at all the points along the line – the field events away from the line do not directly influence this particular particle. So to determine the forces affecting the particle we require a knowledge of all the fields everywhere along the track of the particle.

Fields have their origin in material systems. So we have a two-way situation: particles give rise to fields and particles are influenced by fields. This explains how particles can influence each other. A particle P_1 gives rise to a field F_1 and F_1 then influences a second particle P_2. And P_2 gives rise to a field F_2 which influences P_1. In this way we have a mutual action and reaction between P_1 and P_2.

Next I come to the critical question: How does the field F_1 propagate from P_1? The answer is that from each point on P_1 a cone spreads out, exactly the cone considered in the previous section – the cone along which distances are zero. The field radiated at A propagates along the cone shown in Fig. 5, and the field radiated at B propagates along the corresponding cone. From what was said in the previous section this is the case for light. What is now being said is that the same is true for every field known to physics. These cones are the vehicles of communication whereby material systems interact with each other.

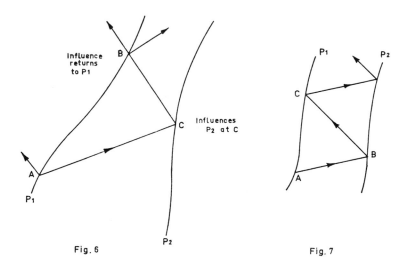

Fig. 6

Fig. 7

Consider how the field radiated by P_1 at A proceeds to influence P_2. This field exists only along the cone. Therefore it can only affect P_2 at the point where the P_2 line cuts the cone, at the point C of Fig. 6. Now because P_2 is influenced – i.e. P_2 is pushed around – the second particle sends out a field along a cone. Where the P_1 line cuts this second cone – i.e., at B – there is now an influence on P_1. What has happened is that P_2 has 'reflected back' the original disturbance from A, and the reflection then produces a reaction on P_1 at B. Starting with the disturbance at A there is indeed the infinite stepladder of backward and forward reflections, shown in Fig. 7.

The same considerations apply to the transmission of information. Suppose the observer P_1 sends out a message from A. Observer P_2, after receiving the message at C, immediately sends his reply which is received by P_1, at B. Interchange of information proceeds along the same stepladder.

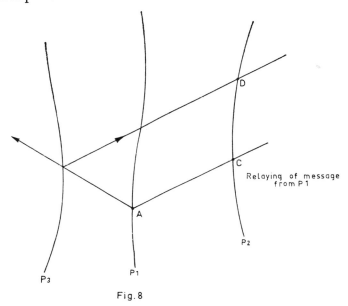

Fig. 8

When more than two particles are considered one must take account of all possible stepladders – there is one between every pair of particles. The problem of interaction or of communication is now far more complex, of course. Dealing with the complexity is a technical problem for the physicist. But one new point of principle is worth mentioning. Consider three observers P_1, P_2, P_3 and let P_1 again send out a message from the point A on his line (Fig. 8). If P_3 were not present there would be only one point on the line of P_2 at which P_2 could receive the message, namely the point at which P_2 cuts the cone from A. However, the presence of P_3 allows the message to be relayed and to be picked up at a second point, D. This leads to the basic

99

physical restriction on communication. A message sent out from A cannot be received by any observer at any point outside the cone from A. The message can, however, be received either on the cone or inside it. For reception inside the cone a relaying station is necessary.

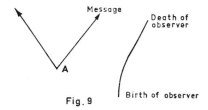

Fig. 9

A deathless observer can always wait to pick up a message — he simply waits until he crosses the cone from the point of transmission. But a practical observer with a line that terminates on death may never reach the relevant cone. There are messages that he can never receive (Fig. 9).

Part II

1. *The arrow of time*

If in ordinary language one interchanges the words 'past' and 'future', statements such as 'I was born in the future and will die in the past' make just as much sense as the more usual 'I was born in the past and will die in the future'. We have simply inverted the meaning of the words. The change is trivial, no different from reading a map from south to north instead of from north to south. Indeed since all times exist with equal reality in our four-dimensional map, the sense in which we elect to read the map is quite irrelevant. When we speak in physics of the arrow of time something much more is implied.

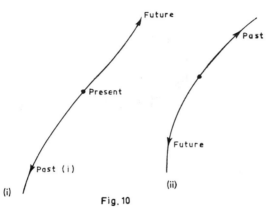

Fig. 10

Consider a particular point on our own personal line in the map. This point we regard as our 'present'. On both sides of the present the line stretches away from us. Which direction we elect to call the future and which the past is unimportant, Fig. 10 (i) and Fig. 10 (ii) are essentially the same – we have simply made a trivial change in the meaning of words. What is not irrelevant is that the events are basically different on the two sides of the present. On one side we are born on the other side we die. This is an absolute difference quite independent of the sense in which we elect to read our map.

(As an aside, we do in fact measure the years backwards up to the birth of Christ. Everybody recognizes this as a mere convention. The convention could easily be extended to dates A.D., by writing them as negative quantities – for example the year 1000 A.D. could equally well be minus 1000 B.C. One can count time backwards without altering actual experience in any way.)

101

But the fact that everyone has birth on one side of the present, death on the other, and that the sense from birth to death *is the same for everyone* is an absolute feature of the world. It is this absolute feature that we call the arrow of time. As we stand at our particular point of the map the territory on the future side is crucially different from that towards the past. Why should this difference exist? This is one of the deepest questions of physics.

A generation ago physicists believed the answer to lie in the science of thermodynamics, but this is not believed today except by those who are unaware of modern developments.

To avoid any further discussion of conventions, suppose we choose the sense of time in the usual way – we were born in the past. Subject to this choice you are given a series of still pictures taken of a set of particles in a box. You are asked to place them in temporal sequence. For the pictures of Fig. 11, the thermodynamicist would give the order (ii), (iii), (i). He would say that the particles were initially in a

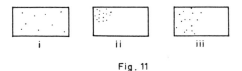

i   ii   iii

Fig. 11

corner of the box and that they spead out until they filled the box. He would point to common experience which shows that particles in a corner tend to spread out, whereas particles that are initially spread out do not tend to go into a corner. And he would then generalize by adding that experience shows material systems tend always towards states of greater uniformity, and to this he would attribute the arrow of time.

This mystique of the thermodynamicist has no chance of surviving against an experienced physicist, however. His bubble is pricked by one question: Is it impossible for the order of the photographs of Fig. 11 to be (i), (iii), (ii) – i.e. the reverse order? After a little technical argument he would be forced to admit that if in picture (i) the particles were made to move in a very special way they could go into the corner, and that (i), (iii), (ii) is a possible order. He would also be forced to admit that the use of the word 'always' in his generalization was incorrect.

The inevitable next step would be for the thermodynamicist to argue that his suggested sequence (ii), (iii), (i) was 'nearly always' right. So the real question becomes – why is he nearly always right?

The answer cannot lie in the laws of physics for the laws of physics permit the opposite sequence, (i), (iii), (ii). The answer can lie only in the way in which the system started off. As a practical matter it is much easier to put the particles in the corner, as in (ii), than it is to arrange all the motions in (i) in the correct way to enable the particles to crowd into the corner. The thermodynamicist is nearly always right just because it is more common for things to start off his way than the opposite way.

By now our friend is in sharp retreat. Instead of the arrow of time depending on some mysterious ordering process among a number of particles the question has been reduced to one of initial conditions. Why are certain types of initial situation much more common than others?

One can come to grips with this question by thinking a little bit more about particles in a box. Suppose that in a particular case the thermodynamicist happens to be correct, and that the order really was (ii), (iii), (i). Suppose that after stage (i) has been reached we take a further series of photographs and we again ask for them to be placed in an order. This time the thermodynamicist will give no answer. He will say that once the particles have become uniformly spread out he can no longer recognize any significant change from one picture to another. He has lost his arrow of time. In fact he only had one so long as the system of particles 'remembered' its initial state – i.e. so long as the particles in the box remembered the situation at the moment the box became closed. If we were to open the box for a moment the system would receive a disturbance. After closing it up again there would be an arrow of time once more, but only so long as the moment of opening was remembered. The thermodynamical arrow of time does not come at all from the physical system itself (which is the thermodynamicist's basic error), is comes *from the connection of the system with the outside world*. It is in the outside world where the arrow is to be found, never inside a closed box. This implies that the arrow of time is derived from the largest scale features of the Universe. Our aim must be to hunt it down.

2. *The asymmetry of light and of force fields in general*

Now that we have seen thermodynamics to be irrelevant to the problem, it is preferable to choose a more profitable point of attack. In Fig. 12 a particle P emits light. According to Part I the light is radiated along a cone. But there are two branches to a cone – the two branches shown in Fig. 13. Why does the light only go out from P

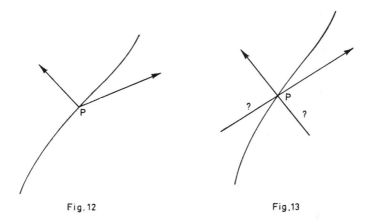

Fig. 12   Fig. 13

in the direction of the arrows? Here we have a crucial asymmetry, far more fundamental than any considerations of thermodynamics.

In Part I I mentioned that order in the world arises from the restrictions that must be satisfied by physical events. Here we have precisely such a restriction. Accepting the usual convention of past and future, light travels from the past to the future, never the opposite way. The same appears to be true for every physical field. I also emphasized that the physical laws are an expression of the restrictions that are found to

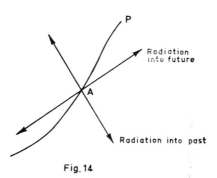

Fig. 14

exist in the world. But just at this stage we have an exceptional and curious situation. Nobody has been able to find laws of physics that forbid the situation of Fig. 14. The physical laws permit a particle to radiate from future to past. Either one must be prepared to accept Fig. 14 as permitted or the physical laws have a very serious incompleteness.

104

I cannot give the technical reasons of symmetry and structure that weigh strongly against the second of these possibilities; suffice it to say that the possibility of revising the laws to permit Fig. 12 but not Fig. 14 appears remote. Rather must one face up to the apparent paradox that while Fig. 14 is possible, only Fig. 12 seems to occur in actuality. The paradox is to be resolved by noticing that Fig. 12 has not really been observed at all, because literally speaking there is only one single particle in Fig. 12, whereas in the Universe there are very many particles. In a many-particle system one must consider the complex stepladder situation discussed in Part I. Perhaps when this is done one can show that light in a many-particle system can travel only from past to future, even though for a single particle it travels both into the future and into the past. Once again we see the arrow of time as residing in the totality of the Universe, not in a local system.

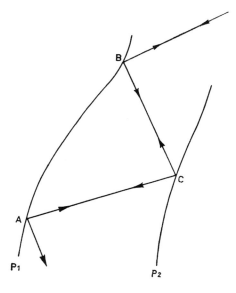

Fig. 15

3. *The work of Wheeler and Feynman*

If particles can radiate into the past as well as into the future, the stepladder of Fig. 7 now goes both ways, as in Fig. 15. If we wish to know what happens as a consequence of some initial disturbance of a particle we have to consider the effects of these reversible stepladders.

105

A disturbance at A of particle P_1 travels to particle P_2 at C where it is reflected. In Fig. 7 the reflection simply carries on into the future, because of the one-way character of the stepladder of Fig. 7. But in Fig. 15 the reflection at C returns an influence back to A. So in order to know the situation in the neighbourhood of the starting point it is not sufficient to deal only with P_1. Reflections back from P_2 must be included; indeed reflections from all the many particles of the Universe must be added along with the direct effect of P_1 itself. Only when a sum total has been arrived at can we decide the full effect of the initial disturbance. Could it be that when all contributions are totalled together we arrive at radiation only into the future? Could it be that the tendency of P_1 itself to radiate into the past is exactly cancelled by the multitude of reflections from all the other particles of the Universe?

This complex problem was tackled about fifteen years ago by Wheeler and Feynman, who found that such a result was a possible consequence of the laws of physics. But there were other possibilities also, and these had to be excluded by starting the whole Universe in a special way. Once things were set going appropriately along an initial time grid line the required result followed everywhere over the whole Universal map.

An interesting feature of this work was the requirement that the Universe be a perfect reflector. Light travelling out into space must encounter sufficient material to become entirely reflected – it may have to travel very far before this happens. The distant regions of the Universe are therefore quite crucial in arriving at the required result, agreeing with one's prognostication from thermodynamical arguments.

4. *The idea of Hogarth*

The necessity of a specially adjusted initial set-up was an unsatisfactory feature of the work of Wheeler and Feynman. It was noticed by Hogarth that Wheeler and Feynman had not taken account of the known fact of expansion of the Universe. Since expansion also implies an arrow of time it was possible that there might be an important effect on the mathematical details of the calculation. This turned out to be true. Hogarth found that when the phenomenon of expansion was included one could dispense with an initial set up. He also found, on reworking the calculations, that of all the cosmologies at present under consideration only the steady state theory – with a continuous creation of matter – gave the required result of radiation into the future but not into the past.

5. Recent work in Cambridge

Because of its cosmological implication I have been much interested in this work. Together with Dr J. V. Narlikar, I have tried to extend the results in several ways. Wheeler and Feynman, and Hogarth also, limited their investigations to the case of light. Narlikar has made similar calculations for other physical fields, and one of them — the neutrino field — confirms very strongly the conclusion that only the steady state cosmology leads to the required asymmetric propagation into the future, never into the past.

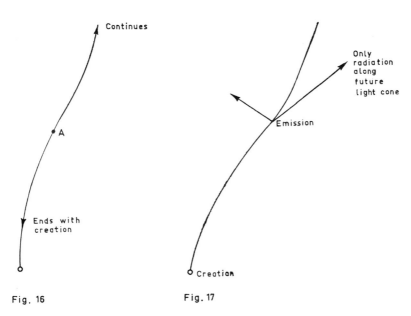

Fig. 16   Fig. 17

6. The arrow of time again

Once again let me remark that by the arrow of time I mean the rooted difference between past and future. We can now describe the arrow by a number of physical phenomena:

(1) Creation of matter. This gives an arrow because a particle of matter possesses a line in the universal map that has an end, but not two ends, as in Fig. 16. If the point A is the 'present', then going from A in one direction we never reach an end (to this direction we attach the convention of 'future'), whereas in the other direction we reach an end.

(2) Expansion of the Universe.

(3) The asymmetric propagation of light and of other physical fields.

(4) Thermodynamical experience.

(5) Subjective experience.

My point now is that these five arrows are interconnected, they are not independent. Given the first, the other four follow inevitably. Given creation of matter as a physical possibility, it is possible to prove by strict mathematics that the Universe must expand, that as time proceeds in the sense of the first arrow distances between the galaxies must increase. Given the second arrow it appears from the work of Wheeler, Feynman, Hogarth, etc. that the third arrow must follow; and given this the fourth follows – it is possible to show that the momentary opening of a closed box containing any physical system in general disturbs it in the manner so dear to our thermodynamical friend. Since our brains are such boxes the fifth arrow follows from the fourth. We can say that if the physical laws are such that matter is created then time's arrow is explained and understood.

7. *Summary*

The locations of physical events in the world require four numbers, three for space, one for time. The events that constitute a particle form a line. If matter is created, the line has one termination but not two. Proceeding along the line, towards the termination is defined as going into the 'past', proceeding in the opposite direction represents going into the future.

Particles interact with each other in such a way that physical fields, the vehicles of communication, are represented by half-cones, not by full cones; when a particle radiates, the radiation propagates along the half-cone that contains the future section of the line of the particle, as in Fig. 17. This asymmetry produces the one-way stepladder system of Fig. 7. It leads to the causal character of the Universe. In particular the realm of experience described by thermodynamics is a consequence of this asymmetry. Thermodynamics has no deeper significance.

It is imperative in the asymmetric system of Fig. 16 that the radiation be considered as a grand total, not only of the direct effect of the particular particle itself but of all the reflections from all other particles. These reflections come from the future. If it were not for reflections from the future the everyday world of commonsense would not exist.

108

New Ideas of Space and Time

P. A. M. Dirac[*]

The Florida State University, Dept. of Physics, Tallahassee, USA

There are reasons for believing that the gravitational constant, expressed in terms of atomic units, is not accurately constant, but varies slowly with the age of the universe. This requires a modification of Einstein's theory of gravitation. One may suppose that the element of distance occurring in Einstein's theory is not the same as that measured by atomic apparatus. It becomes possible to revive Weyl's geometry, a generalization of that used by Einstein, to account for the electromagnetic field. There is then no exact symmetry between positive and negative electric charge, or between past and future.

The standard physical ideas of space and time are based on Einstein's theory of relativity. The special theory requires us to look upon space and time as a four-dimensional continuum, with a geometry differing somewhat from Euclidean geometry, called Minkowski geometry. Einstein's general theory requires physical space to be curved, like the curved spaces dealt with by Riemann in the last century. Such a space can be pictured as a four-dimensional curved surface in a flat space with a larger number of dimensions. The curvature provides a very satisfactory explanation of the gravitational field.

For several decades physicists have accepted this view of space and have used it as the background for interpreting all physical phenomena. However, there are now reasons for suspecting that it may have to be changed.

From the constants of nature one can construct some dimensionless numbers. The important ones are hc/e^2, which is about 137, and the ratio of the mass of the proton to that of the electron, M/m, which is about 1840. There is no explanation for these numbers, but physicists believe that with increasing knowledge an explanation will some day be found.

Another dimensionless number is provided by the ratio of the electric to the gravitational force between an electron and a proton, namely e^2/GMm. This has a value about 2×10^{39}, quite a different order from the previous ones. One wonders how it could ever be explained.

The recession of the spiral nebulae provides an age for the universe of about 2×10^{10} years. If one expresses it in terms of some atomic unit, say e^2/mc^3, one gets a number about 7×10^{39}, which is comparable with the previous large number. It is hard to believe that this is just a coincidence. One suspects that there

is some connection between the two numbers, which will get explained when we have more knowledge of cosmology and of atomic theory.

One can set up a general hypothesis, which we may call "the Large-Numbers Hypothesis", that all dimensionless numbers of this order that turn up in nature are connected. One of these large numbers is the epoch t, the present time reckoned from the time of creation as zero, and this increases with the passage of time. The Large-Numbers Hypothesis now requires that they shall all increase, in proportion to the epoch, so as to maintain the connection between them. One can infer that the gravitational constant G, measured in atomic units, must be decreasing in proportion to t^{-1}.

Now Einstein's theory of gravitation requires that G shall be constant; in fact with a suitable choice of units it is 1. Thus Einstein's theory of gravitation is irreconcilable with the Large-Numbers Hypothesis.

The Large-Numbers Hypothesis is a speculation, not an established fact. It can become established only by direct observation of the variation of G. The effect is not too small to be beyond the capabilities of present-day techniques. Shapiro's (1968) measurements of the distances of the planets by radar are extremely accurate, and if G is really varying to the required extent, it should show up in his observations in a few years time.

We shall assume that the Large-Numbers Hypothesis is correct and the question will be faced as to how Einstein's theory is to be modified to agree with it.

A simple way of effecting a reconciliation is to suppose that the Einstein equations refer to an interval ds_E connecting two neighboring points which is not the same as the interval ds_A measured by atomic apparatus. By taking the ratio of ds_E to ds_A to vary with the epoch we get G varying with the epoch. The ratio is sufficiently nearly constant for the modification in Einstein's theory to be very small.

[*] Lecture held at the Lindau Nobelpreisträger-Tagung, July 2, 1973.

The measurements ordinarily made by physicists in the laboratory use apparatus which is fixed by the atomic properties of matter, so the measurements will refer to the metric ds_A. The metric ds_E cannot be measured directly, but it shows itself up through its influence on the equations of motion. It forms the basis of all dynamical theory, whether the theory is the accurate one of Einstein or the Newtonian approximation. The relation of the two metrics is exemplified by radar observations of the planets. Here a distance which is determined by equations of motion is measured by atomic apparatus.

Let us consider the relation between the two ds's. We must refer again to the Large-Numbers Hypothesis. We bring in another large dimensionless number, the total number of nucleons in the universe. If the universe is infinite, we replace this by the total number of nucleons in the part of the universe that is receding from us with a velocity less than $1/2c$.

This number is rather uncertain because we do not know how much invisible matter there is. Presumably the amount of invisible matter is not very much larger than the visible. We then get a number somewhat around 10^{78}. From the Large-Numbers Hypothesis, this must vary in proportion to t^2. It follows that new protons and neutrons are continually being created.

The question arises: Where are they created? There are various assumptions one might make.

A. Matter is created uniformly throughout space, hence mostly in intergalactic space.

B. Matter is created where it already exists, in proportion to the amount existing there.

C. Matter is created in special places, e.g. in the centers of galaxies.

I do not know which assumption to prefer. In any case the creation of matter is a new physical process, not explicable in terms of any of the known physical processes. The effect is too small to be easily detectable, except possibly with B. This assumption would requiere all matter to multiply, including the matter in the earth. There might be some difficulty in understanding how the matter in very old rocks can have multiplied without disrupting their crystal structure.

Let us return to the relation between ds_E and ds_A. Take as an example the motion of the earth around the sun. According to Newton, the basic formula is

$$GM = v^2 r$$

where M is the mass of the sun, r is the radius of the earth's orbit and v is the velocity. The formula must hold both in Einstein and atomic units. In Einstein units all the quantities are constants. In atomic units they may vary with t. Let us write the formula then

$$G_A M_A = v_A^2 r_A.$$

The velocity v_A is a certain fraction of the velocity of light, the same fraction as in Einstein units, and is thus constant. We have already seen that G_A is proportional to t^{-1}. With assumption A or C, M_A is constant, and so r_A is proportional to t^{-1}. With assumption B, M_A is proportional to t^2 and r_A is proportional to t.

Thus with assumption A or C, the earth is approaching the sun, and the whole solar system is contracting. With B the earth is receding and the whole solar

system is expanding. These effects are cosmological, and are to be superposed on other effects arising from known physical causes. Shapiro's observations should show them up.

The relation between ds_E and ds_A is now

$$ds_E :: t \, ds_A \text{(for A or C)}, \qquad ds_E :: t^{-1} \, ds_A \text{(for B)}.$$

If it is accepted that the Einstein ds is not the same as the atomic ds, it becomes reasonable to make a further development in our ideas of space.

The electromagnetic field is very similar to the gravitational field. They both lead to forces inversely proportional to the square of the distance. This has led people to believe that the electromagnetic field should also be explained as a property of space, instead of being merely something embedded in space. One would then expect physical space to involve a more general geometry than the Riemann geometry of the Einstein theory, leading to both the gravitational and the electromagnetic field in a unified way.

Such a geometry was proposed by Weyl shortly after the appearance of Einstein's theory. Weyl's basic assumption can be easily explained in terms of the idea of parallel displacement.

Given a vector located at a certain point, one can suppose it to be shifted to a neighboring point keeping parallel to itself. One may continue shifting it like this along a path. One may take the path to be a closed loop, so that the vector ends up at the point where it started. If the space is curved the final direction will not be the same as the initial direction.

For a Riemann space, which can be pictured as a curved space immersed in a flat space of a higher number of dimensions, the length of the vector will not change as it is taken round the closed loop. Weyl imagined a more general space in which the length of a vector changes as well as its direction.

Such a generalization is very natural to a mathematician. However, it is not possible for physical space if the lengths of the vectors are referred to atomic standards. The length then cannot change when the vector is shifted. For this reason Weyl's geometry was not accepted by physicists, and Weyl himself gave it up after a few years.

The situation is different if one allows two standards for the measurements of length, ds_A and ds_E. Under parallel displacement the length of a vector referred to ds_A must not change, but it may be that the relation between ds_E and ds_A is not so precise as is given by the above formulas and that the length referred to ds_E does change. One may then have Weyl's geometry applying to ds_E.

If one is working with Weyl's geometry, one cannot talk about the length of a vector without introducing an artificial standard of length at each point and referring the length of the vector to the artificial standard at the point where the vector is situated. When the vector is shifted by parallel displacement through dx^μ, its length l will change by

$$dl = l \varkappa_\mu \, dx^\mu$$

with some coefficients $\varkappa_\mu$. If one changes the artificial standard of length by the factor λ, then it is easily verified that $\varkappa_\mu$ changes according to

$$\varkappa_\mu \rightarrow \varkappa_\mu + \partial (\log \lambda) / \partial x^\mu.$$

110

VI. *The Triumph of Time*. The Triumph of Time, by Pieter Breughel the elder (1574). Father Time occasionally associated with Death and often a malevolent figure in the iconography of art is here shown eating a child (not an unusual association), with Death behind him, while Fame 'blows his trumpet'. In Time's hands is a symbol of eternity (a snake biting its own tail) and he sits on an hour glass with a clock in the tree above him.

This transformation of the $\varkappa_\mu$ must be of no physical significance.

We have just what is needed in order to be able to identify the $\varkappa_\mu$ with the electromagenetic potentials. The geometry is now described by the $g_{\mu\nu}$ of the Einstein theory toge .er with the $\varkappa_\mu$, determining the gravitational and electromagnetic fields in a unified scheme.

The Weyl interpretation of the electromagnetic field as influencing the geometry of space and not merely as something immersed in a Riemannian space has a striking consequence—symmetry-breaking. Consider a charged particle and take a field point P close to its world-line. For simplicity, suppose the coordinate system to be chosen so that the particle is momentarily at rest.

Now take an element of length l at P and suppose it to be shifted by parallel displacement into the future, by an amount δx^0. From the fundamental formula, it will change by

$$\delta l = l\varkappa_0 \,\delta x^0.$$

Here $\varkappa_0$ will consist mainly of the Coulomb potential arising from the charged particle. Suppose the sign of the charge is such that l increases when it is shifted into the future. With the opposite sign of the charge it will decrease. Now there is no symmetry between a quantity increasing and the same quantity decreasing.

Consequently there is no symmetry between positive and negative charge.

If l increases when it is shifted into the future, it decreases when it is shifted into the past. So there is no symmetry between future and past. But if one changes the sign of the charge and also interchanges future and past, one gets back to the original situation.

Atomic physicists have introduced the operators P for changing the parity, C for charge conjugation and T for time reversal. In elementary theories all these symmetries are preserved. The present theory does not provide any breaking of the P symmetry. However, it does break the C and T symmetries, while preserving their product CT.

Experimentally, all three symmetries are observed to be broken, but the product PCT is conserved, so far as is known. It would seem that the breaking of the P symmetry must be ascribed to the short-range atomic forces. However, the breaking of C and of T, with preservation of CT, is caused by the long-range forces, if they are handled in accordance with Weyl's geometry. It would seem that this symmetry-breaking arises from the interaction of the gravitational and electromagnetic fields. The effect must be small, because gravitational effects are always small in the atomic domain.

Received July 5, 1973

Thermodynamics, Cosmology, and the Physical Constants

P.T. Landsberg

ABSTRACT

Certain problems arise when cosmology and thermodynamics are combined, and these are explained. Reasons are given why the cosmological arrow of time should not be regarded as more fundamental than other arrows of time. It is shown by analogy how certain restrictions in classical thermodynamics can be lifted if a self-gravitating object is included in one's thermodynamic system of interest, and if general relativity is used. A matter-plus-radiation type of cosmology is discussed which leads to the view that the universe is old, having gone through many cycles. Finally the values of the constants, h, c, G, and H are used, together with the qualitative assumption that G and H depend similarly on time, to give rough estimates of the mass of an elementary particle, the number of particles in the universe, and the ratio of the gravitational to the electrical force between two particles. The same type of argument can be used to show that if electric charges can depend only on h, c, and H, of which only H depends on time, then these charges must be independent of time.

1. INTRODUCTION

Seven years ago I appended to my discussion before the Society[1] 19 views on irreversibility and entropy, and today I should like to supplement these by adding some

114

physicists' views concerning the arrow of time (Appendix). Contradictions are revealed in both cases, and caution should be induced; all the more so if one considers
the categorical nature of some of the statements cited. My own views have been reformulated in an inaugural lecture last year[2] and I must therefore forego the temp-
tation to try and paint a comprehensive picture of the physicist's time concept on
the present occasion.

Nevertheless some broad viewpoints will be put forward: that certain specific problems arise when cosmology and thermodynamics are combined (§2); that, while cosmology
is important, the view that the cosmological arrow dominates all other arrows of time
is not established (§3); that the inclusion of gravitation and of general relativity
lift certain restrictions imposed by classical thermodynamics (§4); and that there
is some evidence that the universe has gone through many cycles already (§5). The
delight of a theore tical physicist is to deduce quantitative estimates from qualitative assumptions. This is attempted in Section 6 where the number of particles in
the universe and the mass of a typical stable particle are deduced from the numerical
values of h, c, the gravitational constant G and the Hubble parameter H, together
with the qualitative assumption that H and G have the same time dependence.

2. RELATIVISTIC COSMOLOGY + THERMODYNAMICS = SOME DIFFICULTIES

In this section attention will be drawn to two difficulties which arise when one
relates thermodynamics to general relativity and cosmology, difficulties which do
not seem to have been pointed out before. (a) When a system of interacting particles
is considered, the interactions may tend to cancel out in a large system. This occurs
in a plasma consisting of positively and negatively charged particles, since a large
enough volume tends to be electrically neutral. It then hardly interacts electrically
with another large and electrically neutral part of the same system some distance
away. However, gravitational interactions do not cancel, and large but distinct parts
of a system are generally in gravitational interaction. Under these condition the
energy, which in normal thermodynamics is an extensive quantity, loses its exten-

115

sivity; on doubling the system its energy does not double because of the gravitational interactions which enter as an "extra". The same point applies to the entropy. This means that in relativistic cosmology energy and entropy are not extensive: if one halves a volume element, ones needs a convention which tells one how to associate the gravitational interaction energy with each of the two halves. On the other hand, in theoretical analyses one employs the cosmological principle and simple metrics, which asserts precisely that the energy (or entropy) in any comoving volume element is the same as that in any other of the same size. We thus have the puzzle that whereas simple models in cosmology imply extensivity, gravitational interaction denies extensivity, and hence some convention is called for to harmonize the two views. (b) It is well-known in relativity that when the masses of objects are large enough the objects become unstable and collapse. Hence large masses do not necessarily exist. A well-known example is provided by the Chandrasekhar limit: white dwarf stars cannot be more massive than about 1.4 solar masses. This type of consideration is in flat contradiction with a procedure known as taking the "thermodynamic limit". This procedure secures simplified equations for a thermodynamic system by imagining it to become infinitely massive and infinitely extended, while keeping its mass density at the original value. It is widely used. Thus the existence of limiting masses is incompatible with arguments based on the thermodynamic limit.

The above considerations result ultimately from the fact that systems of particles interacting by long-range forces (such as Coulomb or gravitational forces) ought to be ruled out on p. 1 on any book on thermodynamics, even though they seldom are [I did it on p.9 [3]] . Later on systems can be considered which are subject to externally applied fields. Even Coulomb systems can be considered because of the simplifying feature mentioned above. But self-gravitating systems are more difficult to handle. Their gravitational collapse (to black hole states) does not seem to be reproducible in computer studies which are, however, confined to small system. These studies show that such systems can contract by ejecting a particle of comparatively high kinetic energy. Indeed the absence of an equilibrium state makes it difficult to apply thermodynamics unless some dissipative medium is also included in the considerations.[4,5]

116

The way around the difficulty (a) in cosmology is to use relativistic equations which include the dynamic effects of the gravitational interactions in some kind of smeared-out universe. When it comes to the equations of state of the cosmological fluid, the gravitational effects are not imposed again: first of all, it is difficult and little of this sort of theory has been attempted. Secondly, this second introduction of gravitation may involve a kind of "double counting," and should be best avoided. The combination of cosmological equations (which take account of gravitation) and of statistical mechanics (in which gravitation is ignored) leads therefore to hybrid theories. Most present theories are of this type.

We shall adopt this approach also in this paper (see Section 5). We shall get around difficulty (b) by simply not using the thermodynamic limit.

3. THE COSMOLOGICAL ARROW OF TIME DOES NOT DOMINATE

It has been suggested that thermodynamic irreversibility is due to cosmological expansion. It is said that it causes the darkness of the night sky and the red shift of the spectral lines and it is therefore a real phenomenon which can cause other effects such as irreversibility. In a sense cosmology contains all subjects because it is the story of everything, including biology, psychology, and human history. In that single sense it can be said to contain an explanation also of time's arrow. But this is not what is meant by those who advocate the cosmological explanation of irreversibility.[6] They imply that in some way the time arrow of cosmology imposes its sense on the thermodynamic arrow. I wish to disagree with this view.

The explanation assumes that the universe is expanding. While this is current orthodoxy, there is no certainty about it. The red-shifts might be due to quite different causes. For example, when light passes through expanding clouds of gas it will be red-shifted.[7] A large number of such clouds might one day be invoked to explain these red shifts. It seems an odd procedure to attempt to "explain" everyday occurrences, such as the diffusion of milk into coffee, by means of theories of the universe which

117

are themselves less firmly established than the phenomena to be explained. Most people believe in explaining one set of things in terms of others about which they are more certain, and the explanation of normal irreversible phenomena in terms of the cosmological expansion is not in this category.

As regards the darkness of the night sky, it shows indeed that the universe is a sink for radiation, which it absorbs as a gourmet does his paté. Does this *prove* that there is a cosmological arrow of time and that the universe is expanding? Not at all; a static universe with stars of finite ages could also produce a dark night sky, and there are several other possible explanations.[8]

I am not saying that the universe *is* static, but I *am* saying that the theory of the dominance of the cosmological arrow cannot be established by the above arguments. I regard as the most fundamental arrow that which is furnished by statistical thermo-dynamics (See Appendix).

Professor Park has suggested to me a wider interpretation of the notion of a cosmo-logical arrow. The term could be used without caring about what precisely the universe is doing, in particular whether it is expanding or not. The universe is then simply a large system with which every small system is in interaction. While this is indeed another way of using the idea of a cosmological arrow, I would prefer this type of consideration to be considered as part of statistical thermodynamics. The reason is that specific properties of the universe are not used, so that it has really the status of a reservoir.

How the continuing increase of entropy can be explained without reference to the evolution of the universe will be illustrated by means of a picturesque example in the next section.

4. ENTROPY CAN INCREASE WITHOUT LIMIT IN AN ISOLATED SYSTEM WITH GRAVITATION

A gravitational system reaches on contraction states of more and more negative total energy, since the gravitational potential energy of masses m_1, m_2 separated by a distance r is $-Gm_1 m_2/r$, where G is Newton's gravitational constant, and r can approach the value zero. What effect does this have on the entropy?

Although gravitational thermodynamics is only in its early stages, one can see the effect of such systems by recalling some old friends[9] in a Himalayan hilltop community which was completely isolated except that it provided for fuel and food by paying porters who carried these things up the mountain. To pay for this, without reducing the total happiness of the community, it was always arranged that some money ($£x$) was taken from a rich man, and a smaller sum ($£y$) was given to a poorer man in the community. Although y is smaller than x, the poor man's gain in happiness was as great as, or even greater than the rich man's loss in happiness, for two pounds means less to a rich man than one pound means to a poor man. The difference of $x-y$ pounds was available to the porters. Now this community increased in happiness until in the end it was in danger of dying out, since a state was being approached in which all men had the same amount of money and so nothing was available for paying the porters. This model community would have increased in happiness, which was its self-imposed constraint, and then it would have died. This indeed would have corresponded to the heat death of the universe, or indeed to the attainment of equilibrium in any isolated system according to classical thermodynamics. Time's arrow would eventually have died out.

This correspondence to classical thermodynamics is brought about because the entropy of an isolated system never decreases, just as the happiness in our Himalayan community never decreases. Furthermore the final state of a classical physical system is one of uniform temperature (this corresponds to the maximum happiness). Underlying this model is the assumption that money must not be printed in the community, and this corresponds to the conservation of energy according the first law of thermodynamics.

I left this community to linger in its perilous state sixteen years ago. Like the

119

writer of thrillers who cannot afford to let his hero die, I must now relate a
marvellous event. One of the porters, Mr. P. say, hearing of the precarious state of
the community, offered to join it for good. "But", he said, "I shall be different
from all the others, for I shall not mind accumulating any amount of debts." The
social scientists in the community of course did not know how to assess the happiness
of this immigrant, and their situation illuminates our situation as present-day
physicists, in that we have difficulties with the entropy of self-gravitating
systems. However, the porter's action enabled new life to pulse through the community. When the porters who had carried food and fuel up wanted to be paid, Mr. P's a
account was debited. The village printed extra pound notes and gave a little to some
members of the community, while keeping the rest for food and fuel. In this way new
inequalities of wealth began to develop in the community and, although Mr. P's debts
rose to astronomical proportions, he did not mind, and indeed some transactions
became possible again in the community, and this is how it happened.

Every Friday all money was deposited in banks; accounts (all positive) were added up
on Saturdays; and the total W (i.e., wealth in terms of readily available cash) of
the community was determined. This remained constant for long periods while there
was no external trading, since of course money was not being printed or lost. Mr. P's
arrival did not change the constancy of the total liquid cash W. But it did mean that
by assigning a debit balance to his account, the community was able to print money
which could be assigned to certain accounts without altering the total W. This increased happiness and set up inequalities which then formed a basis for renewed
trade inside the community. The total happiness *did not decrease* in this trade and
money for food and fuel could again be put aside. When the stage of equal wealth
and hence stagnation approached again, Mr. P's account was debited by an additional
amount and an equal amount of money was printed. Without changing W, trade inside
the community was again able to flourish. In this way the community was able to live
with increasing total happiness for ever after.

The source of this renewal was Mr. P's account which could go increasingly into the

red, just as gravitational potential energy becomes more negative as a body contracts.[†] This is the first consequence of note, and it is connected with the fact
that general relativity does not know a law of energy conservation and that the
integrated proper energy of the cosmological fluid does not have to remain constant
but can change. The manner of this change arises from the details of the processes
involved (this has not been modelled here). Thus the proper energy of an element of
the cosmological fluid decreases with time during expansion, increases during contraction, and in both cases its magnitude is affected by the pressure [equation
(5.2), below].

As a second consequence note that there is no final limit on the happiness which can
be generated, once the money supply can in principle be increased. It illustrates
that in relativity a developing model of the universe which has some irreversibility
built into it (i.e., happiness is increasing rather than remaining constant) is able
to exhibit an ever increasing integrated proper entropy. It does not reach a final
state of maximum entropy compatible with the given total energy available at the
time, as expected from classical thermodynamics. This is due to the presence of
gravitation in a system, consisting of two or more fluids[10, 11].

These ideas should help with the next section.

5. HAS THE UNIVERSE ALREADY GONE THROUGH MANY CYCLES?

Considering a homogenous and isotropic Friedmann model with matter and radiation, one
knows that if they are independent of each other each develops adiabatically, i.e.,
according to the law

$$T_i V^{\gamma_i - 1} = A_i' \qquad (i = m \text{ or } r)$$

where V is a standard comoving volume, T_m and T_r are the temperatures of non-relativistic matter and radiation, the A_i' are constants, and

---

[†]Quantum effects may provide a lower limit, in which case Mr. P's account can carry
only finite debts.

121

$$\gamma_m = \frac{5}{3} \qquad \gamma_r = \frac{4}{3}$$

Now $V \propto R^3(t)$ where R is the cosmological scale factor which is a function of time. It follows that, if the A_i are other constants,

$$T_m R^2 = A_m , \quad T_r R = A_r , \quad T_m/T_r = \frac{A_m/A_r}{R} \qquad (5.1)$$

Suppose now that the interaction energy rises without limit as $R \to 0$. It is then reasonable to suppose that matter and radiation are in equilibrium, in the sense that $T_m = T_r$, at times near the big bang. As expansion proceeds, and R rises, the interaction weakens and one would expect the matter temperature to drop more rapidly $(T_m \propto R^{-2})$ than the radiation temperature $(T_r \propto R^{-1})$. During the long period including the present there is weak interaction and (5.1) holds. If there is an oscillation, compression will turn non-relativistic matter into relativistic matter which also satisfies $T_m R = $ constant, so that $T_m R$ should drop during this period. During high compression at the end of the cycle the strong interaction will make the matter and the radiation temperature again roughly the same. Indeed, as matter gains heat from the still hotter radiation, one may even find that T_m/T_r can overshoot unity for a while, but there should then follow a return to equilibrium.

Similar arguments should hold for the total energies U_m, U_r. During the middle part of the cycle one would expect the standard results

$$U_m \propto R^{-3} , \qquad U_r \propto R^{-4}$$

Thus the energy of the radiation would drop more rapidly during the expansion and rise more rapidly during the contraction.

If E_m is the rate of transfer of energy from radiation to matter, the entropy production rate as a function of time can be taken to be

$$\dot{S} = \left[\frac{1}{T_m} - \frac{1}{T_r}\right] E_m$$

This is always non-negative, provided only

$$E_m > 0 \rightarrow T_r > T_m \qquad \text{energy passes to matter}$$

and

$$E_m < 0 \rightarrow T_r < T_m \qquad \text{energy passes from matter}$$

The constraint $\dot{S} > 0$ can be realised, for example, by choosing the radiative inter-action

$$E_m = A U_m U_r (T_r^4 - T_m^4)/(\text{volume})$$

when entropy is found to increase continuously,[12] and the other statements made above can also be verified.

At high compression the entropy is roughly constant, since there is almost equilibrium. For purposes of machine computation one can cut out the singularity and start the next cycle with the same entropy at the same values of U_m, U_r, and R as occurred at the cutoff, but with an equal and opposite value of dR/dt, which turns the contraction into an expansion. This is of course a highly irregular procedure since such singularities as the contraction of the universe to a point follow from the equations of motion if they are taken literally. This singularities have been the subject of much work, some of which is described in this volume by Dr. P. C. W. Davies. Nonetheless there are some (and the present author is one of them) who regard such "point states" of the universe as unphysical. This is not to deny the importance and intrinsic interest of the singularities. It is, however, a justification for the highly irregular procedure to which reference has just been made.

123

If U_o is a fixed unit of energy and R_o a fixed value of the scale factor, one can use dimensionless variables

$$y_m \equiv U_m/U_o, \quad y_r \equiv U_r/U_o, \quad r \equiv R/R_o$$

At the maximum extent of the computed model one finds the values given in Table 1. The total energy at maximum extent is seen to increase with cycle number while the energy density at maximum extent decreases with cycle number.

Table 1 -- Values at the Point of Maximum Expansion

	y_m	y_r	$y \equiv y_m + y_r$	r	y/r^3
1st cycle	1.032	0.158	1.190	2.697	0.06066
2nd cycle	1.032	0.161	1.193	2.704	0.06034
3rd cycle	1.032	0.164	1.196	2.711	0.06003

This may be understood in terms of the basic equations of the model, which are

$$\dot{U} + 4\pi p\, R^2 \dot{R} = 0 \qquad (p = p_m + p_r, \quad U = U_m + U_r) \tag{5.2}$$

$$\tfrac{1}{2}\dot{R}^2 + \frac{G}{C^2}\frac{U}{R} = C$$

where C is a negative constant and p is the pressure. If the maximum of R in cycle j is denoted by $R^{(j)}$, it follows that, since U and R have opposite signs, U goes through its minimum when R goes through its maximum. Also at this maximum

$$U^{(j)} \propto R^{(j)}, \quad \text{i.e.} \quad \rho^{(j)} \propto [R^{(j)}]^{-2}$$

where ρ is the total energy density. The model shows that $R^{(j)}$ (i.e., $r^{(j)}$) increases with j, hence $\rho^{(j)}$ (i.e., $y^{(j)}/r^{(j)3}$) decreases with j. Thus, away from the endpoints, the energy density drops to lower and lower values as the cycles follow each other.

124

This makes it difficult for someone alive in this model during a "late" cycle to believe that there is an adequate energy density present for another contraction to occur. This seems to be the precise position in which we find ourselves. The model thus suggests an explanation of the fact that cosmologists find it difficult to decide if the universe will contract again, or if it will continue to expand indefinitely.

As the cycles continue the entropy goes on increasing, and the model is then found to make full use of the freedom bestowed on a thermodynamic system which includes gravitation and is relativistic: its entropy can increase without limit (see Section 4).

Our model is somewhat artificial and so we did not pursue its properties to many cycles. Its properties are those already expected on general grounds, but it furnishes a specific example of these general principles and shows how one can calculate through the discontinuity. It furthermore finially disproves the notion that the entropy law excludes a cyclic world process.[13]

The model also suggests the possibility that the cycles may have started off with a minor fluctuation which have become larger and larger as the millennia passed. It is rather like a weight oscillating at the end of a spring which becomes weaker as time passes, thus allowing the oscillations to become larger.

6. AN ESTIMATE OF THE MAIN MASSES IN THE UNIVERSE

In this concluding section I should like to tell a different story. It belongs to a meeting on *Time* because the crucial assumption to be made is that Newton's gravitational constant G depends on time, and that it does so in precisely the way in which the Hubble parameter H depends on time. This hypothesis, due to Dirac, is forty years old and the justification is as follows.

The ratio of the electrostatic to the gravitational forces between proton and electron

125

is a huge number

$$\frac{e^2/r}{Gm_p m_e/r} \approx 10^{40}$$

e = electron
p = proton

Let the age of the Universe, or more precisely the time since the last big bang, be T. From the reciprocal of the present value H_o of the Hubble parameter it is

$$T \approx H_o^{-1} \approx 10^{17} \text{sec}$$

One then finds that T expressed in terms of an atomic unit of time such as $e^2/m_e c^3 \approx 10^{-23}$ sec is also of order 10^{40}. An empirical relation results of the type (other units of time and other masses could be chosen)

$$G \approx \frac{2^4}{Tm_e^2 m_p c^3}$$

Now the rough equality of two distinct huge dimensionless numbers is perhaps no accident, but could be fundamental, and therefore valid at *all* times. This "large number hypothesis" implies a time-dependence of G such that if c and the masses of the elementary particles are assumed constant,

$$G(T) \propto H(T)$$

After this preliminary remark we proceed with description of more recent work.

In cosmological theories which involve as the only fundamental quantities G, H, and the constants c and h, all masses must have the form $h^\alpha H^\beta G^\gamma c^\delta$, whence dimensional analysis shows that the unknown constants α, β, γ, δ, can all be reduced to one unknown constant which will be called b. The other constants (e, h, c) will be assumed time-independent, in agreement with current views.[14] One then finds that all masses are reducible to the basic masses

126

$$m(b) \;=\; k(b) \left[\frac{h^3 H}{G^2}\right]^{1/5} \quad \left[\frac{c^5}{hH^2 G}\right]^{b/15} \qquad\qquad (6.1)$$

where $k(b)$ are unidentified dimensionless constants. Dimensional analysis depends on the assumption that these constants are of order unity, and this will be assumed.

The point about making the Dirac hypothesis is that $m(b)$ will in general be time-dependent unless the value of b is such that $m(b)$ depends only on the ratio H/G. In other words the hypothesis taken together with (6.1) leads to a unique value of b which should furnish time-independent masses, i.e. masses of the stable elementary particles. Now in terms of time dependence

$$m(b) \;\propto\; H^{(3-2b)/15} \Big/ G^{(6+b)/15} \;\propto\; G^{-(b+1)/5}$$

Hence $b \simeq -1$ for elementary particles. The interactions between particles, and other effects, make this only a rough theory, but Fig. 1 shows that the rest masses of the known particles do cluster around the value $b = -1$. If in fact G decreases with time,

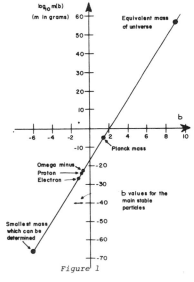

Figure 1

127

then the large masses $m(b)$ increase ($b > -1$), the small masses decrease ($b < -1$), while length standards based on $b = -1$ remain constant in time. A typical example of the latter is the Compton wavelength $h/cm(-1)$.

The smallest mass determinable in the period $T \simeq H^{-1}$ since the last big bang is from the uncertainty relation $h/c^2 T$ which is of the order $m(-5)$, and it should become smaller with lapse of time, while the mass equivalent of the energy of the universe is of order

$$m(9) \simeq c^3/GH \qquad (6.2)$$

which increases with time.[15]

In order that this analysis be compatible with an oscillating universe the time periods involved must be confined to the interval between the beginning of the last cycle and the period of maximum extent of that cycle. In this way one can avoid the infinite mass (6.2) at the maximum extent, when $H = 0$. It would appear therefore that Dirac's hypothesis can be valid only within limited periods in each cycle if an oscillating universe is assumed.

If one grants this, one can take several further steps. One of them is to estimate the number of particles in the universe at

$$n \simeq \frac{m(9)}{m(-1)} \simeq \left[\frac{c^5}{hGH^2} \right]^{2/3} \simeq 24 \times 10^{80}$$

As regards the time dependence of n since the last big bang, we have

$$n \propto G^{-2}$$

128

because $H \propto G$. We may ask: Why is the dimensionless quantity n so large, contrary to the usual properties of dimensionless numbers? The answer could be: Because G has decreased so much since the last big bang; it is large because the universe is old[16], or, more precisely, it has come a long way since the last big bang.

The ratio λ of the gravitational to the electric forces between two particles can now be calculated, for a time-independent electronic charge e. One has, if the fine structure constant is denoted by $\alpha \equiv e^2/hc \approx 1/137$,

$$\lambda = \frac{Gm^2/r^2}{e^2/r^2} \approx \frac{1}{\alpha} \frac{Gm(-1)^2}{hc} = \frac{1}{\alpha} \left[\frac{hH^2 G}{c^5} \right]^{1/3} = \frac{1}{\alpha \sqrt{n}} \approx 10^{-43}$$

The smallness of the ratio λ is thus a consequence of the theory outlined, and depends on the length of time since the last big bang. The other cosmological coincidences can be explained in the same way.[15]

There is some evidence for the time-dependence of G, which is crucial to the present argument[17,18]: $\dot{G}/G \approx -8 \times 10^{-11}$ (years)$^{-1}$. Such variations of G can be incorporated in Newtonian[19] and relativistic[20,21] theory.

As an admirer of F. Hund, I want to amend his remark (See Appendix) that past and future are markedly different "because the universe is young." An entropy curve is still steeply rising with time in the early part of a cycle and then flattens out near the maximum. In this region the difference between past and future is liable to be blurred. We now incorporate this view into our philosophy by remarking that the strong difference between past and future is due to the fact that in the present cycle we are not yet close to the epoch of maximum extent. There is now no reference to an absolute beginning of the history of the universe, as is implied by Hund's reference to its youth. I insist only that the universe is at an early stage of its cycle, even though this may be the hundredth cycle of an old and tired universe. These comments may illustrate the remark, attributed to Bondi,[22] that the universe has a low entropy because "the night sky is very black with very bright points, the stars, in it": the entropy may be relatively low even in an old but oscillating universe.

There has been some discussion in the literature concerning the possibility that the basic electronic charge depends on time, but the consensus is that it does not. One can investigate this point by dimensional analysis. The coupling parameter for the interaction is now e^2 instead of Gm^2, and we seek to determine its order of magnitude. Suppose it can depend on h, H, and c. Then the possible dependence on time is implied by the possible dependence of e on H. Let us put

$$e = kh^{\alpha}H^{\beta}c^{\gamma}$$

where k is a dimensionless constant. Then

$$\left[M^{\frac{1}{2}}L^{3/2}T^{-1} \right] = \left[M^{\alpha}L^{2\alpha+\gamma}T^{-\alpha-\beta-\gamma} \right]$$

and the resulting three equations imply

$$\alpha = \gamma = \tfrac{1}{2}, \qquad \beta = 0$$

so that

$$e = k(hc)^{\frac{1}{2}}$$

Thus e is independent of time and k^2 is in fact equal to the famous fine-structure constant $e^2/hc \approx 1/137$.

REFERENCES

1. P. T. Landsberg, 1970, "Time in statistical physics and in special relativity," *The Study of Time,* Vol. 1, p. 59.

2. P. T. Landsberg, 1975, *A Matter of Time* (Southampton University). An inaugural lecture.

3. P. T. Landsberg, 1961, *Thermodynamics with Quantum Statistical Illustrations* (New York: Wiley).

4. R. H. Miller, 1973, "On the 'thermodynamics' of self-gravitating N-body systems," Ap. J. *180*, 759-782.

5. M. J. Haggerty and G. Severne, 1974, "Monotonic evolution of Boltzmann's H in weekly coupled gravitational fields," Nature *249*, 537-538.

6. B. Gal-Or, 1972, "The crisis about the origin of irreversibility and time anisotropy," Science *176*, 11-17.

7. P. T. Landsberg, 1971, unpublished.

8. P. T. Landsberg and D. A. Evans, 1972, "What Olbers might have said," in *The Emerging Universe* (Ed. W. C. Saslaw and K. C. Jacobs; Charlottesville: University Press of Virginia) p. 107-130.

9. P. T. Landsberg, 1961, *Entropy and the Unity of Knowledge* (Cardiff: University of Wales Press). An inaugural lecture.

10. R. C. Tolman, 1934, *Relativity, Thermodynamics and Cosmology* (Oxford: University Press).

11. P. C. W. Davies, 1974, *The Physics of Time Asymmetry* (Leighton Buzzard: Surrey University Press; Berkeley: University of California Press).

12. P. T. Landsberg and D. Park, 1975, "Entropy in an oscillating universe," Proc. R. Soc. *A346*, 485-495.

13. E. T. Whittaker, 1942, *The Beginning and End of the World* (Oxford University Press) p. 39. The Riddell Memorial Lectures.

14. A. M. Wolfe, R. L. Brown and M. S. Roberts, Phys. Rev. Letts. *37*, 179 (1976).

15. P. T. Landsberg and N. T. Bishop, 1975, "A Cosmological deduction of the order of magnitude of an elementary particle mass and of the cosmological coincidences," Physics Lett. *53A*, 109-110.

16. S. Weinberg, 1972, *Gravitation and Cosmology* (New York: Wiley), p. 621.

17. T. C. van Flandern, 1975, "A determination of the rate of change of G," M. N. Roy. Astr. Soc. *170*, 333-342.

18. D. S. Dearborn and D. N. Schramm, 1974, "Limits on the variation of G from clusters of galaxies," Nature *247*, 441.

19. P. T. Landsberg and N. T. Bishop, 1975, "A principle of impotence allowing for Newtonian cosmologies with a time-dependent gravitational constant," M. N. Roy. Astr. Soc. *171*, 279-286.

20. C. Brans and R. H. Dicke, 1961, "Mach's principle and a Relativistic Theory of Gravitation," Phys. Rev. *124*, 925-935.

21. N. T. Bishop, 1976, "Cosmology and a general scalar-tensor theory of gravity," (Unpublished).

22. J. Gribbin, 1975, "Oscillating Universe bounces back," Nature *259*, 15-16.

APPENDIX

Remarks made by physicists on the direction of time

The selection is somewhat arbitrary and possibly not typical of the cited authors'

present views.

1. Statistical thermodynamics furnishes the clue

"Only the second law of thermodynamics indicates clearly
a direction of time".
C. F. V. Weizsäcker, 1939.

"It is not at present clear if...it is possible to deduce
the law of entropy increase from classical mechanics".
L. D. Landau and E. M. Lifshitz, 1966.

"We thus come to an interesting relation between the effect
of the cosmological arrow of time and of the 'microscopic arrow of
time' on the thermodynamic evolution of a system: both suppress the
anti-kinetic behaviour and lead to an irreversible approach to
equilibrium".
A. Aharony, 1971.

"No asymmetry between the two directions of time is to be
found in the general laws of nature; it is due to another fact about
the world, namely its low entropy. The first appearance of the earlier
states of still lower entropy cannot be understood physically.....
Past and future are so markedly different because the universe is still
very young".
F. Hund, 1972.

"If, however, the choice were between abandoning the
Friedmann models and deriving the direction of time from some source
other than cosmology (such as thermodynamics) then I think most
physicists would chose the latter".
G. J. Suggett, 1975.

"Suppose a system develops without interference from the
outside. Then it chooses among its available equilibrium states
in proportion to their realisabilities. I shall call this principle *P*.
Time has a direction in virtue of principle *P*, but, for more
penetrating observers, the direction of time has to be *derived* by
averaging".
P. T. Landsberg, 1975.

2. Statistics does not provide the answer

"If all the laws of physics are time-symmetrical, they
would not be able to describe a contracting universe".
"Surely it is not by rejecting information about our system
that we can make it reveal to us the sense of time which it would not
otherwise show".
T. Gold, 1958.

"It is somewhat offensive to our thought to suggest that if
we know a system in detail then we cannot tell which way time is going,
but if we take a blurred view, a statistical view of it, that is to
say throw away some information, then we can..."
                                        H. Bondi, 1962.

        "The 'arrow' of time...does not seem to be of a stochastic
character".
                                        K. Popper, 1965.

3. Explanations using past and future as unexplained (or primitive) concepts

        "But while the distinction of prior and subsequent events
may be immaterial with respect to mathematical fictions, it is quite
otherwise with respect to the events in the real world. It should
not be forgotten...that while the probabilities of subsequent events
may often be determined from the probabilities of prior events, it is
rarely the case that probabilities of prior events can be determined
from those of subsequent events, for we are rarely justified in
excluding the consideration of the antecedent probability of the prior
events".
                                        J. W. Gibbs, 1902.

        "We can now understand the anisotropy of time. The future
is, by definition, the direction in which prediction is possible....
A complete mathematical description of the universe must unfold from
a description of the initial state. It is not possible to reconstruct
the past history of the universe by working backward from a complete
macroscopic description of the present state".
                                        D. Layzer, 1967.

        "One can *define* a direction of time by stating that the
instant of reception is *later* than the instant of emission...this point
of view makes it possible to avoid the paradoxes and confusion found in
the literature. In fact, the invariance of the mechanical laws for
time reversal is not contradictory with the distinction between past and
future just introduced".
                                        L. Rosenfeld, 1972.
                                        See also in Caldirola, 1961, page 3 and
                                        in Gold 1967, page 193.

        "Irreversibility and the generalised second law are derivable
from the existence of...two categories of instants: an 'information
gathering category' (the past), and a 'prediction category' (the future).
The existence of these two categories seems to be a fundamental feature
of nature, not explainable in terms of the second law or in terms of
any other physical law".
                                        A. Hobson, 1971.

4. Gravitation is responsible

        "...a falling apple a kilometer away over an arc of ten
centimeters is ample to mix up the trajectory of a mole of normal gas
in a time of milliseconds..."

        "...the arrow of time then only an illusion? It is the
purpose of this note to answer stoutly the arrow is real, that is,
not subjective, that it is not essentially cosmological, that it arises
from an inescapable feature of all physical theory".
                                        P. Morrison, 1966.

                            135

"We have reached a remarkable conclusion. The origin of
all thermodynamic irreversibility in the real universe depends
ultimately on gravitation. *Any* gravitating universe that can exist
and contains more than one type of interacting material *must* be
asymmetric in time, both globally in its motion, and locally in its
thermodynamics".
P. C. W. Davies, 1974.

5. The undiscovered half of the universe is the culprit

"One finds that the physical difference between the two
directions of time can be explained only by the circumstance that there
are in the world areas which do not satisfy the theories [classical
and wave mechanics] which lead to thermodynamics".
M. Bronstein and L. Landau, 1933.

"In a world that was perfectly symmetric with respect to the
time axis, matter of opposite time senses would decouple...all
observers would be aware of an apparent asymmetry in time...that was
merely a measure of their ignorance of the other half of the universe".
F. R. Stannard, 1966.

6. Other contenders: Quantum mechanics, electrodynamics, boundary conditions,
    and interactions with the surroundings

"The two directions of time are not equivalent in quantum
mechanics and it is possible that the law of entropy increase is the
'macroscopic' description of this state of affairs".
L. D. Landau and E. M. Lifshitz, 1966.

"It is sometimes said that electrodynamic theory itself
introduces time asymmetry...it is a matter of experience that
retarded potentials give the correct answer while advanced do not.
Is this a case of time's arrow being contained in each elementary
process... ? More careful consideration shows that this is not so".
T. Gold, 1974.

"...the laws of physics are symmetric in regard to time,
and so the asymmetry must arise from boundary conditions".
W. H. McCrea, 1975.

"The thermodynamic arrow of time does not come at all from
the physical system itself..., it comes from the connection of the
system with the outside world".

"We can say that if the physical laws are such that matter
is created then time's arrow is explained and understood".
F. Hoyle, 1962.

"It is entirely possible and consistent to speak of the
atypically behaving branch systems, whose entropy increases are *counter
directed* with respect to those of the majority, as *decreasing* their
entropies in the positive direction of time...we are able to give the
usual temporal description of fluctuation phenomena in this way..."
A. Grünbaum, 1974.

REFERENCES TO THE APPENDIX

Aharony, A., 1971, "Time reversal, symmetry violation and the H-theorem." Physics
        Letters, *37A*, 45-46.
Bondi, H., 1962, "Physics and Cosmology." Observatory, *82*, 133-143. (Halley Lecture).
Bronstein, M. and Landau, L., 1933, "On the second law of thermodynamics and global
        connections in the cosmos." Phys. Z.d. Soviet Union, *4*, 114-119. In German.

Caldirola, P. (Ed.), 1961, *Ergodic theories*. (New York: Academic Press.)

Davies, P. C. W., 1974, *The Physics of Time Asymmetry*. (London: Surrey University Press) page 109.

Fraser, J. T., Haber, F. C. and Müller, G. H. (Ed.), 1972, *The Study of Time*. (Berlin: Springer.)

Gal-Or, B. (Ed.), 1974, *Modern Developments in Thermodynamics*. (New York: Wiley.)

Gibbs, J. W., 1902, *Elementary Principles in Statistical Mechanics*. (New Haven: Yale University Press) page 150-151.

Gold, T., 1958, at the 11th International Physics Congress, Solvay. See also Am. J. Phys. *30*, 403-410, (1962) Richtmyer Lecture).

Gold, T. (Ed.), 1967, *The Nature of Time*. (Ithaca: Cornell University Press.)

Gold, T., 1974, "The world map and the apparent flow of time," in Gal-Or (Ed.), 1974, 63-72.

Grünbaum, A., 1974, "Popper's view on the arrow of time," in *The Philosophy of Karl Popper* II (Ed. P. A. Schlipp). (LaSalle, Illinois: Open Court) page 793.

Hobson, A., 1971, *Concepts in Statistical Mechanics*. (New York: Gordon and Breach) page 156.

Hoyle, F., 1965, "The Asymmetry of Time," Third Annual Lecture to the Research Students' Association, Canberra, 1962 (Canberra: Australian National University).

Hoyle, F. and Narlikar, J. V., 1974, *Action at a Distance in Physics and Cosmology*. (San Francisco: Freeman.)

Hund, F., 1972, "Time as physical concept," in: Fraser, J. T., et al. (Eds.) 1972, 39-52. In German.

Landau, L. D. and Lifshitz, E. M., 1966, *Statistische Physik* (Berlin: Akademie Verlag). Translation of the second Russian edition, revised by E. M. Lifshitz.

Landsberg, P. T., 1975, *A Matter of Time*. (Southampton: University) 18-19. An inaugural lecture.

Layzer, D., 1967, "A unified approach to cosmology," in *Relativity and Astrophysics* (Ed. J. Ehlers), Lectures in Applied Mathematics, Vol. 8, (Providence, R. I.: Am. Math. Soc. Press).

McCrea, W. H., 1975, review of P. C. W. Davies, 1974, in Nature, *253*, 485.

Morrison, P., 1966, "Time's arrow and external perturbations," in *Preludes in Theoretical Physics* (Ed. A. de-Shalit, H. Feshbach, L. van Hove). (Amsterdam: North-Holland) 347-351.

Popper, K., 1965, "Time's arrow and entropy." Nature, *207*, 233-234.

Rosenfeld, L., 1972, "General introduction to irreversibility." In *Irreversibility in the Many-Body Problem* (Ed. J. Biel and J. Rae). (New York: Plenum), page 10.

Stannard, R. F., 1966, "Symmetry of the time axis." Nature, *211*, 693-694.

Suggett, G. J., 1975, review of Hoyle and Narlikar, 1974. Nature, *254*, page 223.

von Weizsäcker, C. F., 1939, "The second law and the distinction between past and future." Ann. d. Phys., *36*, 275-283. In German.

Part C

Quantum Mechanics including Black Holes

VII.   *The need to be down to earth.* This illustration shows a vehicle which registers the distance travelled—a 'way-measurer' or hodometer. It was undoubtedly a precursor of the mechanical clock (figure I) as it relied on a system of toothed wheels constituting a reduction gear train. It activates slowly, revolving pins which release catches at regular intervals. This in turn enables the figures at the top of the carriage to strike a drum or gong. The particular instrument illustrated here comes from the Han dynasty and is a rubbing of a stone-cut from a tomb shrine of AD128.

142

Volume 37A, number 1   PHYSICS LETTERS   25 October 1971

TIME REVERSAL SYMMETRY VIOLATION AND THE H-THEOREM

A. AHARONY

Department of Physics and Astronomy, Tel-Aviv University, Tel-Aviv, Israel

Received 11 September 1971

Computer experiments with a dilute gas of hard disks with a time reversal non invariant inter-action are described. The effect on the H-function and on the anti-kinetic behaviour is discussed.

The recent interpretation of data related with the decay of the neutral kaon indicates that time reversal symmetry T is violated in this decay [1]. Since all usual treatments of thermodynam-ical irreversibility assume that the microscopic equations of motion are symmetrical under time reversal, it now becomes interesting to abandon this assumption, and search for macroscopic thermodynamical effects of the microscopic time reversal symmetry violation.

For obtaining a preliminary qualitative feeling on the behaviour of a gas of particles interacting with a time reversal violating potential, the method of computer experiments has been used. For a general velocity dependent potential, the force between each pair of particles has compo-

nents in both the directions of r and v (the rel-ative distance and velocity of the two particles). As the simplest possibility for representing such an interaction, a simple extension of the hard sphere model was used, as follows: The parti-cles move on straight lines, unless two of them approach a distance σ from each other. At this point, their relative velocity is instantly chang-ed, to the form:

$$v \rightarrow v - \alpha \frac{(r \cdot v)}{r^2} r - \beta v \tag{1}$$

with

$$\alpha = (1-\beta)\left[1 + \left(1 + \frac{r^2 v^2}{(r \cdot v)^2} \cdot \frac{\beta(2-\beta)}{(1+\beta)^2}\right)^{1/2}\right] \tag{2}$$

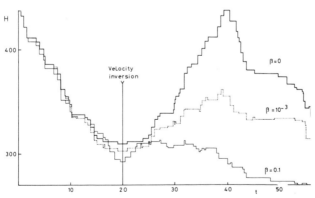

Fig. 1. The time dependence of the Boltzmann H-function, for 100 particles in a 2-dimensional box, with the time reversal violation parameter $\beta=0$, 10^{-1}, 10^{-3}.

Clearly, the case $\beta = 0$ $(\alpha = 2)$ corresponds to the usual hard sphere collision. The term βv corresponds to the time reversal violating part in the potential. Thus β may be considered as a time reversal violation. The expression for α, eq. (2), ensures conservation of energy.

The algorithm for calculating the motion of the particles is similar to that of Alder and Wainwright [2], including only the hard core part. The boundaries are taken as hard walls. In the present calculation, 100 particles were run in a two dimensional box. At $t = 0$, the particles were put on the vertices of a square mesh in the box, with velocities of equal magnitude and arbitrary directions. The hard core radius, σ, was taken as 0.01, the mesh unit as 1, and the magnitude of each particles initial velocity as 1. The Boltzmann H-function was calculated after each collision, with the Alder-Wainwright expression

$$H = \sum_i n(v_i^2) \left[\ln n(v_i^2) - \ln(4\pi v_i^2 \Delta v_i) \right] \tag{3}$$

with

$\Delta v_i = 0.1$.

After $t_0 = 20$ (around 30 collisions), all velocities were inverted, and the calculation continued. This kind of inversion has been discussed by Balescu [3] analitically, for a simple exponential potential, and by Orban and Bellemans [4], in a computer experiment similar to ours. The purpose of this inversion is to enable us to examine the "anti-kinetic" evolution of the gas, due to the correlation created at the time t_0.

Fig. 1 describes the time dependence of the Boltzmann H-function of the system, with several values of β. The case with $\beta = 0$ follows the same pattern as in Orban and Bellemans' fig. 1(a) [4]; after t_0 the H-function retraces its path exactly back to its initial value, and then starts to decrease again. In the case with $\beta = 0.1$, near-

ly no anti-kinetic stage is observed; the system nearly immediately "forgets" the correlations it had at t_0, and continues to approach equilibrium. The case with $\beta = 10^{-3}$ is intermediate, with H still increasing after t_0, but not as fast or as high as for $\beta = 0$.

These results have an interesting resemblance to those of Orban and Bellemans, who obtained a suppression of the anti-kinetic behaviour by introducing random errors. A similar result of Alder and Wainwright's is mentioned by Morrison [5], who relates this suppression with the cosmological arrow of time: no system is really isolated except the whole universe, and every finite system is slightly perturbed by its interaction with the whole universe. This perturbation is simulated by the random errors. We thus come to an interesting relation between the effect of the cosmological arrow of time and of the "microscopic arrow of time" on the thermodynamical evolution of a system: both suppress the anti-kinetic behaviour and lead to an irreversible approach to equilibrium.

In all three cases the system approaches the Maxwell-Boltzmann distribution, but a larger value of β corresponds to a faster decrease of H. Thus, the H-theorem seems to hold even "more strongly" when a time reversal non invariant interaction is introduced.

References
[1] Y. Achiman, Nuovo Cimento Letters 2 (1969) 301;
    R. S. Casella, Phys. Rev. Letters 21 (1968) 1128;
    22 (1969) 554.
[2] B. J. Alder and T. E. Wainwright, J. Chem. Phys.
    31 (1959) 459; Nuovo Cimento Suppl. 9 (1958) 116.
[3] R. Balescu, Physica 36 (1967) 433.
[4] J. Orban and A. Bellemans, Phys. Letters 24A
    (1967) 620.
[5] P. Morrison, Preludes in theoretical physics, eds.
    A. de Shalit et al. (North-Holland, 1966) p. 347.

* * * * *

Time Reversal Asymmetry at the Fundamental Level — and its Reflection on the Problem of the Arrow of Time

Y. NE'EMAN

Tel Aviv University, Ramat Aviv, Israel
The University of Texas, Austin, U.S.A.

Theoretical physicists are humbled by the realization that almost 150 years after the promulgation of the Second Law of Thermodynamics, they cannot yet consider the Law and its supposed connection with the arrow of time as a solved problem. This is due to several twentieth century developments, some of which we shall list here, gradually focusing our attention on recent work relating to the breakdown of the hypothesis of time-symmetry at the fundamental level.

a) Thermodynamics represents a direct, phenomenological, macroscopic discipline. Statistical Mechanics was invented to provide it with a foundation at the microscopical level. Within that framework, the H-theorem has indeed been provided as a microscopic basis for the Second Law.* However, the connection with the Arrow of Time is not reproduced: under time-reversal, entropy is seen to increase (or the H function to decrease) in both time directions [1]!

b) Two additional effects have been discovered in which there appears an arrow of time, with no clear a priori correlation with the thermodynamical arrow: the expansion of the universe (1929) and the time-reversal asymmetry [2] in neutral kaon decays (1964). It was further proved experimentally that there exists no direct causal connection between these two, i.e. that the microscopic CP violation is not caused by a long range "cosmological" field [3] (though there still exists one loophole, a superweakly coupled spinless boson with two units of strangeness [4]).

c) An apparent clash between the Second Law and biological evolution was apparently removed in the thirties and the interplay of these dissipative systems with the non-biological background was explained. We still lack some good quantitative examples. Some workers in the field have further claimed that the same mechanism could also provide an explanation for the physiological time-arrow, and thus connect the latter to the Second Law [5]. Though invoking the techniques

---

* *Editor's note*: Yet statistical mechanics fails to deduce the origin of irreversibility. Its origin is not to be found in the mathematics of statistical mechanics

of the Mathematical Theory of Information, no convincing proof of these claims has been provided to-date, and the question of the physiological arrow is still open.

d) The cosmological time-asymmetry is still not entirely understood, mainly because of the uncertainty with respect to the actual physical cosmological model. It is nevertheless clear that a very wide class of solutions (probably including the physical one) allows for both expansion and contraction at different periods (even the steady-state theory possesses a time-reversed ever-contracting case, in which creation of matter is replaced by annihilation). It is claimed by many workers in the field that the correlation between thermodynamics and cosmology is such as to invert the thermodynamical arrow in the contraction phase, so that only expansion is realized (this is indeed the main theme of this volume, and we dispense with references; points of view may differ, going all the way from a search for consistency to an attempt to use cosmology as a "pointer" in thermodynamics [6]).

e) In a recent publication, we have proved [7] that this is allowed as far as the microscopic arrow is concerned, provided *CPT* symmetry holds; a redefinition of which is matter and which is antimatter is implied.

f) The proofs used in Statistical Mechanics for the *H*-theorem involved an assumption of time-reversal symmetry at the fundamental level. Aharony [8] has reviewed the situation now that this can be assumed no more (see article in this volume). In most cases, unitarity is sufficient, but the Boltzmann equation's extension to quantum statistics by Uehling and Uhlenbeck does not hold [9].

g) Similarly, the Onsager relations are violated to first order in ε, the time asymmetry parameter [10].

h) The role of the microscopic asymmetry has been further studied through the derivation of an appropriate Master equation [11]. The existence of an *H* theorem is ensured by randomness in the baths describing the kaon decay products. The method was used to prove the existence of an oscillating contribution to the *H* function, due to the *T* violating interaction: this new function represents the additional irreversibility it generates.

i) It should be noted that the microscopic *T* violation could thus also provide a "pointer" [6], either in addition to cosmology or replacing it. However, we know as yet of no quantitative treatment which would serve as a criterion for the choice of such a pointer [12]. Remembering that the issue in question consists in the removal of one half of the statistically allowed evolutions for the *H* function (i.e. allowing it to decrease in one direction only), the entire search for pointers might be a misplaced one, the answer lying in an improved use of information theory and perhaps in the elucidation of the physiological correlation. In that case, both cosmological and microscopic arrows would still have to be consistent with the solution, but they would not have to be shown as large enough to produce the entire observed irreversibility in nature.

j) Note that some of the complexity in our problem may be due to the fact that we have all been dealing with a topic lying astride the otherwise clear-cut division between laws of nature and boundary conditions. Except for the microscopic arrow, the laws might all have been time-symmetric, with the asymmetry due to boundary conditions only. This seems impossible now that we have found a T-asymmetric interaction.

Paragraphs e, f, g, h represent a concrete advance in the understanding of the role of the CP violation beyond the scope of conventional Elementary Particle Physics. Hopefully, we shall also find an exact answer to the possibility mentioned at the end of paragraph b. It seems to us however that any further serious advance will depend upon progress in the application of information theory methods to answer the following questions:

1) Is there a way of allowing the H function to decrease in one direction only, by taking into account the actual method by which we acquire information about the world, and without the help of an external factor (a "pointer")?

2) Does this also explain the physiological arrow?

Depending upon the answer to the first question, we shall know more about what to look for in cosmology or in kaon decays, relating to the arrow of time.

Note Added in Proof

A further "arrow" is introduced at the fundamental level, through the postulation of micro-causality and of the vacuum as the lowest energy state. These are of the nature of the second law in that they do not break T-symmetry fundamentally, their role consisting only in ensuring consistency in the time progression in one direction, after this has been picked by the boundary conditions .

REFERENCES

[1] See for example the discussion in LANDAU-LIFSCHITZ' *Statistical Physics*, § 8. For an enlightening numerical example, see ORBAN, J. and BELLEMANS, A., *Phys. Letters* **24A**, 620 (1967).
The argument can now also be formulated in the phenomenological theory of thermodynamics.
[2] CHRISTENSON, J. H., CRONIN, J. W., FITCH, V. L. and TURLAY, R., *Phys. Rev. Lett.* **13**, 138 (1964).
The possible implications for the Arrow of Time problem were raised by G. ZWEIG, Conference on Decays of K-mesons, Princeton-Pennsylvania Accelerator, Nov. 1967.
NE'EMAN, Y., March 1968 session of the Israel Academy of Sciences (available as No. 13 in the *Proceedings of the Section of Sciences*).
The experimental proof that T itself is indeed violated is presented in ACHIMAN, Y., *Nuovo Cimento Letters* **2**, 301 (1969); CASELLA, R. S., *Phys. Rev. Lett.* **21**, 1128 (1968) and **22**, 554 (1969).

[3] DE BOUARD, X., et al., *Phys. Letters* **15**, 58 (1965); GALBRAITH, W., et al., *Phys. Rev. Lett.* **14**, 383 (1965).

See also remarks on p. 539 of the review by LEE, T. D. and WU, C. S., *Ann. Rev. Nucl. Phys.* **16**, 511 (1967).

[4] GÜRSEY, F. and PAIS, A., unpublished. It is felt (though without proof) that even this $|\Delta S| = 2$ field can be experimentally excluded (or that its range can be shown to be finite) by studying the exchange of two quanta, a $|\Delta S| = 0$ contribution acting between any two nucleons, for instance (but with a $R^{-n}, n \geq 3$ dependence).

[5] See for example, COSTA DE BEAUREGARD, O., *Logic, Methodology and Philosophy of Science* (Y. BAR-HILLEL, ed.), North-Holland, Amsterdam 1965, 313.

This does represent an advanced attempt in a very promising direction, but is still rather qualitative. One hopes that further advance along this line will indeed answer the questions we raise in our concluding remarks.

[6] MORRISON, P., *Preludes in Theoretical Physics* (A. DE SHALIT et al., eds.), North-Holland, Amsterdam 1966,347.

[7] AHARONY, A. and NE'EMAN, Y., *Int. J. Theor. Phys.* **3**, 457 (1970).

[8] AHARONY, A., article in this volume.

See also AHARONY, A. and NE'EMAN, Y., *Proceedings of the 1972 Coral Gables Conference on Fundamental Interactions* (to be published).

[9] This is analyzed in Ref. 8. It had already been noted by TER-HAAR, D., *Revs. Mod. Phys.* **27**, 289 (1955).

[10] AHARONY, A., to be published.

The results are included in the second paper of Ref. 8.

[11] AHARONY, A., *Ann. Phys.* **67**, 1 (1971) and **68**, 163 (1971).

[12] AHARONY, A., *Phys. Lett.* **37A**, 45 (1971).

Looking at the graph in this article we would say that a 10 % T-violation does generate an overall impression of irreversibility, whereas the 0.1% case does not.

THE "ARROW OF TIME" AND
QUANTUM MECHANICS   A. J. Leggett

In those exciting but frustrating fields of knowledge, or perhaps one should say ignorance, where physics tangles with philosophy, the difficulties usually lie less in finding answers to well-posed questions than in formulating the fruitful questions in the first place. The attempts which follow to do this for one particular area could charitably be described as at best quarter-baked, and may well reflect the ignorance and confusion of the author rather than that of the scientific community as a whole.

Most natural scientists probably have a deeply ingrained belief that it should be possible to give a complete description of the laws of nature without explicit reference to human consciousness or human intervention. Yet at the heart of physics — for long the paradigm for natural science — lie two problems where this assumption is still subject to furious debate. One is the question of the "arrow of time" or more correctly the apparent asymmetry of physical processes with respect to time,[1, 2] the other the problem of measurement in quantum mechanics.[3] In this essay I speculate on the relationship between these at first sight disconnected problems and the vast areas of human ignorance which may possibly lurk behind them.

At the level of classical mechanics, electrodynamics and (with one minor proviso which is unimportant for present purposes) elementary particle physics, the laws of physics as currently accepted indicate no "preferred" direction of physical processes with respect to time: in technical language, they are invariant under the operation of time reversal. Crudely speaking, every process in (say) mechanics which is possible in one direction is also possible in the reverse direction: for example, if a film were taken of a system obeying the laws of Newtonian mechanics without dissipation (such as is constituted, to a very good approximation, by the planets circling the Sun), and if such a film were then projected backwards, it would be impossible to tell this from the film alone. Moreover — a related but not identical point — the laws of physics at this level give no support to the idea that the present (or past) "determines" the future rather than vice versa. Indeed, while it is certainly true that from a knowledge of the positions and velocities of the planets at some initial time we can in principle calculate the values of the same quantities at a later time, the converse is equally true: from the values at a later instant we can in principle calculate the earlier values just as accurately. Thus, within the framework of mechanics alone, the idea that the earlier events "cause" later ones rather than vice versa is a (possibly illegitimate) importation of anthropomorphic concepts into the subject: and the same is true for electrodynamics and elementary particle theory.†

Any extrapolation of these ideas to the whole of physics would, however, obviously run violently counter to common sense. It is a matter of observation (not of interpretation!) that there exist very many spontaneously occurring processes in nature whose time inverses do *not* occur spontaneously, for instance the melting of ice in a glass of warm water or the gradual loss of energy of a bouncing ball. In such cases a film of the process run backwards would be immediately recognizable as such. Moreover, most people, at least among those brought up in the modern Western intellectual tradition, would certainly say that the present (and past) can influence the future but not vice versa. Perhaps, however, it is worth noting even at this stage that this has not always been the dominant belief and may indeed even now not be the dominant one on a world scale: and while, historically speaking, these societies whose beliefs could be crudely called teleological or fatalistic have not usually embraced natural science in

---

† It is a misconception (though one surprisingly widespread among physicists) that experiments in elementary particle physics have "proved causality" in a sense which would determine a unique direction of the causal relationship in time. On close inspection it turns out (as usual in such cases) that in the interpretation of the experiments the "direction of causality" is already implicitly assumed.

the sense in which we understand it, it is worth asking whether this correlation is a necessary one.

Let us return for the moment to the occurrence in nature of "irreversible" processes, that is, those whose time inverses do not occur spontaneously, and which could therefore apparently be used to define a unique "direction" of time. Such processes are, of course, part of the subject matter of thermodynamics and statistical mechanics, and the conventional explanation of the apparent asymmetry despite the time-symmetry of the underlying microscopic laws goes, very crudely, as follows: if a system is left to itself, its degree of disorder (technically, its entropy) tends to increase as a function of time. To use an often-quoted analogy, if we shuffle a pack of cards we will almost always make it more disordered: if we start with the cards arranged in "perfect order" (ace, king, queen of spades on the top, etc.) we will almost inevitably end up, after shuffling, at a less ordered distribution, but it is extremely unlikely that starting from a "random" pack we would end up, by the ordinary process of shuffling, at a perfectly ordered distribution, and indeed any player who achieved such a result would almost automatically be suspected of cheating. Such an increase of disorder (entropy) seems at first sight to be naturally asymmetric in time and hence to define a unique "direction".†

Let us assume for the moment that the ascription of more "order" to the unshuffled ("perfect") distribution than the "random" one is an intuitively transparent operation which involves no implicit anthropomorphic elements and moreover that it can be made satisfactorily quantitative (cf. previous footnote). Actually, the first of these assumptions at least is by no means unproblematical (what is a pack of cards *for*? Would Martians recognize the perfectly ordered pack as such? etc.) but it is not what I want to discuss here. Rather I want to focus on a difficulty which is well known and has been discussed very thoroughly in the literature,[1, 2] namely that in any purely physical process, the underlying dynamics of the system is time-reversible as mentioned earlier, and therefore that disorder will tend to increase in *both* directions: in fact, if we *knew* for certain that a given pack of cards had been shuffled "at random" and nevertheless found that at the time of observation it was completely ordered, we could legitimately conclude not only that a few minutes later, as a result of further shuffling, it would be more disordered, but also that a few minutes *before* the observation it had also been less ordered, and that we just happened to have caught it at the peak of a very unlikely statistical fluctuation. Now in practice, of course, if we found a perfectly ordered pack *of whose history we knew nothing*, we would conclude nothing of the sort, but rather that it had been deliberately prepared in order by purposeful human intervention. Thus, the apparent asymmetry of the increase of disorder with time in a case of this kind is a consequence of the fact that human beings can "prepare" highly ordered states, whereas they cannot "retropare" them: that is, they can set the initial conditions for a system at time $t < t_1$, and then let the laws of physics take their course undisturbed between, say, t_1 and t_2. If the initial conditions correspond to a highly ordered state, the degree of disorder is practically bound to increase in time despite the time-symmetry of the underlying laws. The inverse procedure is impossible, at least according to common sense: to determine the state of the system at the final instant, t_2, we would have to intervene *before* t_2, that is precisely during the interval when the laws of physics were supposed to be taking their course without outside intervention. So, in the end,

---

† The analogy as presented is, of course, very much oversimplified. In fact, if the "disorder" or "entropy" of a pack of cards is to correspond to the concept used in statistical mechanics, it should be a characteristic not of a particular arrangement of the cards ("microstate") but of a class of arrangements ("macrostate"), having some gross property or properties in common, e.g. a particular value of the number of pairs of neighbouring cards which are of the same colour. For details see, for example, ref. 4. However, for the limited purpose for which I need the analogy here it is unnecessary to go into these complications.

the apparent asymmetry implied by the second law of thermodynamics (increase of disorder) turns out to be intimately related to the fact that we can only affect the future — that is, to the second paradox indicated above, namely the fact that while we have a strong sense of the "direction" of causality in the macroscopic world, microscopic physics supplies no obvious basis for such an idea. Perhaps it is now becoming apparent in what way human consciousness, and human intervention, may at first sight at least be involved in the problem of time asymmetry even within the realm of physics.

At this stage it is as well to digress a moment to dispose of one matter which is not directly relevant to the present argument. It goes without saying that there are plenty of irreversible processes in nature where no human intervention is or could possibly have been involved. In such cases the asymmetry in time is generally and plausibly ascribed to the obvious asymmetry in the natural environment which arises from the fact that the Sun is radiating energy outwards rather than sucking it in "from infinity" — technically, it is a source rather than a sink of radiation. This fact is in turn usually related to the so-called "cosmological" time asymmetry — the fact that (according to most current theories) the Universe as a whole is expanding rather than contracting. This is a topic which itself involves some fairly deep conceptual problems,[8] but it does not directly affect the argument as presented above. However, we shall encounter it again later.

Turning back to the question of situations involving human agency, we have just seen that the apparent asymmetry of such situations in time is a consequence of the inability of human beings to affect the past. The question I now want to raise is: Does physics itself, directly or indirectly, via, say, biological considerations ultimately based on physics, provide any obvious reason for this inability? Or, indeed, for our somewhat related inability to "remember the future"? Such questions might seem at first sight absurd, and indeed may be made so by any of a number of quite natural misinterpretations: if, for instance, we were to interpret the second question as asking why it is the past we remember *rather than* the future, we risk inviting a reply in purely linguistic terms, namely that the past is, by linguistic convention, the "direction" which we can remember. To focus the discussion, therefore, it may be helpful to ask two more specific questions: Does physics by itself forbid: (a) the hypothesis that intelligent beings in a distant part of our Universe (Daleks for short) should have a sense of the "direction" of time which is reversed with respect to ours, (b) a limited degree of genuine precognition among us ordinary human beings?

Hypothesis (a) is only interesting (and indeed perhaps only meaningful) if, apart from their possibly inverted sense of time, the nature of Dalek "life" is sufficiently close to our own to be recognizable as such.† In that case we should have to ask, what are the conditions on our own environment which (for instance) allow an organism to differentiate in one direction in time rather than the other: if, as is commonly accepted, a necessary condition is the constant input of radiation energy from the Sun, then our "biological" and hence our (overall) "psychological" arrow of time is a consequence of the "cosmological" arrow. Since the Daleks are inhabitants of the same universe, they share this latter arrow and their sun is also presumably a source rather than a sink of radiation. It would then follow that hypothesis (a) is excluded — although it is by no means trivial to fill in the details of the above argument.[6]

However, even if we accept this conclusion it is by no means obvious (to me at least) that we should then exclude hypothesis (b) without further discussion. Indeed, to an unprejudiced observer, the evidence,[7] anecdotal as it inevitably is, for a strictly limited degree of

† The question of whether, and how, we could recognize "time-inverted" beings as intelligent[5] is a fascinating one, though unlikely to be of practical interest to space biologists!

precognition may be thought quite impressive. (Ability to "affect the past" would presumably be considerably more difficult to recognize!) And what I want to suggest is, that at least in the absence of a much more detailed understanding of the workings of the human brain than we at present possess, it is *not* entirely obvious that the laws of physics, even when combined with the given *overall* direction of biological process, exclude any possibility of genuine precognition over fairly small distances in time — or, by the same token, of a very limited ability to "affect the past". Needless to say, such a possibility, were it to exist, would have profound implications not only for philosophy but also for our view of physics itself.

Let us now turn to the second, at first sight unrelated, area of physics where human consciousness is sometimes believed to play a special role, namely the theory of the measurement process in quantum mechanics.[3] In the standard formulation of quantum mechanics,[8] one talks strictly speaking not about a single system, but about an "ensemble" (class) of identically prepared systems, and describes such an ensemble by a "wave function", knowledge of which enables us to predict, *purely statistically*, the probability of various outcomes of a given measurement. Consider, for instance, the measurement of some quantity A which can take one of a finite number of discrete values† $a_1, a_2, \ldots, a_i, \ldots$. The wave function then allows us to predict unambiguously the *probability* p_i of getting a specified outcome a_i of the measurement on a system taken at random from the ensemble. Yet the wave-function description certainly cannot be replaced by a description in which a fraction p_i of the systems forming the ensemble are said to be *in* the state corresponding to a value a_i of the quantity A: the two descriptions would in fact give in general quite different predictions of the outcome of an experiment in which a quantity different from A is measured, and to date at least the predictions of the wave function description seems to be in good argument with experiment while the predictions of the rival description are not.

One is therefore apparently forced to say that until the quantity A is measured it *does not have a definite value* for any individual system in the ensemble. (It should be emphasized, again, that it is not just a question of us not knowing the value: if this were the case, the description rejected above would be adequate.) On the other hand, according to the standard prescriptions of quantum mechanics, the moment that the quantity A is measured the description of the system undergoes a discontinuous change: in fact the wave function changes abruptly so as to accommodate the information that A now "has" the value we have just measured (the so-called "reduction of the wave packet"). One is then faced with a dilemma: *either* (a) the wave function characterizes some physical property (which we at present find difficult to interpret intuitively) of each individual system: *or* (b) it is merely a shorthand for the statistical properties of the ensemble (and thus, naturally, has to be rewritten as soon as we obtain additional information about any particular system).‡ The difficulty with interpretation (a) is that in all but the very simplest cases it is hard to see how the assumed physical property can suddenly and discontinuously change when a measurement is made;[9] for instance, if a single electron (or photon) is diffracted through a narrow slit, its wave function spreads out over a wide area — and so, presumably, there is actually some kind of (presently unknown) physical disturbance over the whole of this area. Yet the moment that the position of the particle is "measured" (e.g. by observing the highly localized flash it makes on a scintillating screen) the

---

† Such a quantity might be, for example, the projection of the particle's intrinsic angular momentum on a given axis. Of course, the very fact that only discrete values of such a quantity are allowed is a consequence of the quantum theory and would not occur in classical mechanics.

‡ Technically speaking, by obtaining more information about a given system we assign it to a new ensemble.

wave function is supposed to contract into a very small volume near the position of the flash; and so, presumably, does the associated physical disturbance. Indeed, if we take the prescription literally, the contraction should actually take place at a rate exceeding the speed of light, thereby violating the canons of special relativity. With interpretation (b) above, in which the wave function represents nothing physical but is simply a statistical device resulting from our ignorance, the difficulty is precisely to see why the statistical description rejected above does not work. Faced with this dilemma, the majority of physicists have embraced, at least implicitly, the so-called Copenhagen interpretation[10] (more correctly, non-interpretation) of quantum mechanics according to which it is meaningless to ask questions, awkward or otherwise, about the meaning of the wave function, since it is simply and solely a mathematical interpolation enabling us to infer from a recipe for the preparation of a particular ensemble to the probability of various outcomes of possible experiments on it.[11]

While most practising physicists find this (non)-interpretation quite satisfactory in the context of their everyday use of quantum mechanics, it has become increasingly recognized over the last 20 years or so that the awkward conceptual problems it raises are not going to be exorcised so easily. Indeed, probably the single issue which divides physicists most deeply at present is the extent to which these problems cast doubt on the claims of quantum mechanics to constitute in some sense the ultimate description of physical reality. Let me sketch very briefly the nature of just one of these problems.[3] The notion of "measurement" clearly plays a special role in quantum mechanics, since the wave function is supposed to change continuously and causally between measurements but to jump (collapse) discontinuously and acausally as soon as a measurement is made.[12] This prescription, however, is ambiguous, since quantum mechanics contains no instructions for deciding exactly when a "measurement" has been made. In fact, it is possible to argue that what we normally call a measurement is nothing more than interaction with a (usually man-made) device such as a Geiger counter which itself consists of atoms and molecules, in which case it should itself be described by a quantum mechanical wave function; but if so, the device itself will in general not possess a definite value of all its physical quantities (including the readings of dials, etc.) until these are themselves "measured", and so on in an infinite regress. Indeed, some authors [13] have argued that the only satisfactory way to terminate this regress is to allow the "measurement" to be made only when the reading of the dials (say) is registered by a human mind, thus introducing human consciousness into the theory as an extra-physical ingredient. Clearly the paradox here is somewhat reminiscent of the one we encountered earlier in the context of time-reversibility: to enable thermodynamics to provide a unique "direction of time" we were forced in the end to assume that human beings can affect the future but not the past, an assumption which (apparently) cannot be justified within physics itself: to formulate quantum mechanics at all, we have to introduce the idea of measurement, a notion which (apparently) cannot be defined without internal inconsistency in physical terms alone.

In the face of this and other difficulties, some physicists have speculated that quantum mechanics may actually only be an approximate description of reality, the true description involving a consideration of a sub-quantum level characterized by some variables which at present we are unable to detect experimentally (the so-called "hidden variables"). Such a description, it is hoped, could be completely causal in character, and would if suitably constructed reproduce the quantum-mechanical results for existing types of experiment while avoiding the conceptual difficulties concerned with the problem of measurement; moreover, it might in principle predict results different from quantum mechanics under conditions more stringent than those hitherto attained.[14] A model of such a theory has been explicitly constructed[15] for a simple situation involving spin measurements on a single particle (or ensemble of such), thus finally disproving the widespread but erroneous belief[16] that no hidden-variable theory could reproduce all the results of quantum mechanics even for such a simple system. At first sight, therefore, it looks as if there are no very fundamental *a priori* objections to the idea of such a "hidden-variable" theory.

However, one of the most interesting and surprising developments in fundamental physics in the last decade or so has been the demonstration that the position is quite otherwise as soon as one considers a slightly more complicated situation, namely spin measurement (or the equivalent) on two systems which have interacted in the past but are now very far apart in space (so that, within the framework of presently accepted ideas, they should not influence one another). Let us assume (A) that the hidden-variable theory is such that after the two systems separate, each separately has a state described by its own hidden variables (which we do not, of course, know and which may be strongly correlated, in the usual statistical sense, with those of the other, remote, system); and moreover, (B) that the outcome of any measurement on (say) system 1 is determined only by the state of that system and by the properties of the apparatus set up to perform measurements on it, but not (for instance) by the variables of system 2 or the properties of *its* associated apparatus. (Such a hypothesis seems entirely natural since the two systems, and the two associated pieces of apparatus, are very distant from one another in space.) A theory having these properties is referred to as a "local" hidden-variable theory. There is now a remarkable theorem[17] which states that *no local hidden-variable theory can reproduce all the results of quantum mechanics.* Moreover, with some trivial modifications, the theorem can be strengthened to apply more generally to *any* "local" theory, that is any theory for which assumption (B) above is true and in which (A) is replaced by the more general assumption that there exists a description of the state of each system separately after they have ceased to interact. (In quantum mechanics no such description exists, which itself leads to a well-known paradox.[18] This theoretical conclusion has now been complemented by an experiment[19] which deliberately provided a situation in which the quantum mechanical predictions could not be reproduced by any "local" theory: the quantum predictions were, nevertheless, found experimentally, thus demonstrating fairly conclusively that no "local" description of nature can be correct.

It should be stressed that this conclusion at first sight runs violently counter to common sense, since it says that (at least in certain circumstances) a system cannot even be *described* individually and in isolation, even though it may be spatially separated from all other matter in the Universe. Indeed, if we were to take the argument to its logical conclusion it would seem to say that we can *never* describe any system in isolation, since it must have interacted with something in the past, however long ago!

It is possible to avoid this somewhat unpalatable conclusion if one is prepared to modify one or two of the "common-sense" assumptions embodied, perhaps implicitly, in (A) and (B) above in a sufficiently radical way. For instance, a possible hypothesis[20] is that by setting up the apparatus designed to measure the properties of system 1 one may, in some way not at present understood, affect the physical conditions prevailing in the region where system 2 is and thus the results of measurements made on it. Such a hypothesis, if it is not to violate the special theory of relativity, could in principle be tested by setting up the apparatus "at the last minute", so that there is no time for any signal to be transmitted to system 2.

However, to my mind a more intriguing possibility, and one which at last makes the promised contact with the first part of this essay, is that the "direction of causality" might in some sense be violated in this type of experiment[21] (and, perhaps, more generally in quantum measurement processes). In other words, instead of regarding the initial state of the system (whether described by hidden variables or not) as determining the outcome of measurements made on it, we might regard the outcome of the measurements as determining, at least partly, the initial state. This is perhaps somewhat more plausible in a hidden-variable picture (or some other theory which seeks a "sub-quantum-mechanical" level of reality) since a (temporary) "backward" interpretation of causality at the sub-quantum level might not necessarily conflict with the usual "forward" interpretation of the level of quantum mechanics, nor produce results which are clearly incompatible with the known initial conditions. It should be noticed, by the way, that the typical times involved in experiments of this type are usually extremely small by macroscopic standards (usually of the order of 10^{-9} sec) and indeed are probably less than the shortest microscopic "relaxation time" for the irreversible processes likely to be relevant here. It is therefore perhaps not so unthinkable that phenomena of a type (at first sight) unknown on the time-scale appropriate to the macroscopic world might exist over such short intervals.[22]

Speculating even more wildly, one might hope that if anything remotely resembling this proposal were to be true, it would not only resolve the "measurement paradoxes" of quantum mechanics at the atomic level, but also provide the microscopic basis for those phenomena, if they really exist, which *do* violate the "sense of time" on a *macroscopic* time-scale — such as precognition. That the time scale involved here is so many orders of magnitude longer (perhaps minutes or hours) is perhaps not so strange if one considers that the human brain, regarded as a physical system, certainly possesses a degree of complexity very many orders of magnitude greater than any of the instruments used in physics. On the other hand, it could equally well be that the questions of microscopic and macroscopic violations of the sense of time may turn out to be essentially unrelated, as seems to be the case with the violations of left-right symmetry at the levels of elementary particles and of biology.[23]

The above speculations may seem to be (and no doubt are) vague to the point of irresponsibility. Nevertheless, I do strongly suspect that if in the year 2075 physicists look back on us poor quantum-mechanics-besotted idiots of the twentieth century with pity and head-shaking, an essential ingredient in their new picture of the Universe will be a quite new and to us unforeseeable approach to the concept of time: and that to them our current ideas about the asymmetry of nature with respect to time will appear as naïve as do to us the notions of nineteenth-century physics about simultaneity.†

ACKNOWLEDGEMENT

I am indebted to Dr. Paul Davies for a very helpful discussion and criticisms of the original manuscript.

† For a deeper discussion of many of the questions raised here, as well as many recent references, see O. Costa de Beauregard, *Foundations of Physics* (in press).

1. For general discussions of the problem of time asymmetry, see H. Reichenbach, *The Direction of Time*, University of California Press, Berkeley, 1971, and
2. P. C. W. Davies, *The Physics of Time Asymmetry*, Surrey University Press, London, 1974.
3. For a general introduction to the problem of measurement in quantum mechanics, see B. d'Espagnat, *Conceptual Foundations of Quantum Mechanics*, 2nd edition, Benjamin, New York, 1976.
4. R. Kubo, *Statistical Mechanics*, North-Holland, Amsterdam, 1967, chap. 1.
5. Cf. N. Wiener, *Cybernetics*, Wiley & Sons, New York, 1948, p. 34; *The Nature of Time*, edited by T. Gold, Cornell University Press, Ithaca, N.Y., 1967, pp. 140-2.
6. Cf. H. F. Blum, *Time's Arrow and Evolution*, Princeton University Press, Princeton, N.J., 1968.
7. See, for example, J. B. Rhine, *The Reach of the Mind*, Faber & Faber, London, 1956, chap. 5.
8. L. Eisenbud, *Conceptual Foundations of Quantum Mechanics*, van Nostrand Reinhold, New York, 1971.
9. This point is emphasized in (e.g.) L. E. Ballentine, *Rev. Mod. Phys.* **42**, 358 (1970).
10. N. Bohr, *Phys. Rev.* **48**, 696 (1935).
11. For a more sophisticated version of the "Copenhagen interpretation" see H. Reichenbach, *Philosophic Foundations of Quantum Mechanics*, University of California Press, Berkeley, 1944.
12. For an alternative interpretation which avoids this "collapse" (but involves other difficulties) see H. Everett III, *Rev. Mod. Phys.* **29**, 454 (1957).
13. F. London and E. Bauer, *La Théorie de la mésure en mécanique quantique*, Hermann et Cie, Paris, 1939; E. P. Wigner, *Am. J. Phys.* **31**, 1 (1963).
14. For a forceful statement of this point of view, see D. Bohm, *Causality and Chance in Modern Physics*, Routledge and Kegan Paul, London, 1957.
15. D. Bohm and J. Bub, *Rev. Mod. Phys.* **38**, 453 (1966).
16. This belief goes back to a result of J. von Neumann, *Mathematical Foundations of Quantum Mechanics*, Princeton University Press, Princeton, N.J., 1955.
17. J. S. Bell, *Physics,* **1**, 195 (1964-5).
18. A. Einstein, B. Podolsky and N. Rosen, *Phys. Rev.* **47**, 777 (1935).
19. S. J. Freedman and J. F. Clauser, *Phys. Rev. Letters* **28**, 938 (1972).
20. D. Bohm in D. R. Bates (ed.) *Quantum Theory*, vol. III, Academic Press, New York, 1962, p. 385.
21. To the best of my knowledge this possibility was first pointed out in the present context by O. Costa de Beauregard, *Revue Internationale de Philosophie*, no. 61-62, 1 (1962); *Dialectica* **19**, 280 (1965).
22. Cf. the problem of "pre-acceleration" in electrodynamics (Ref. 2, p. 125), where the relevant time is of order 10^{-23} sec.
23. But see T. L. V. Ulbricht, *Q. Rev. Chem. Soc.* **13**, 48 (1959).

Singularities and Time-asymmetry—an excerpt
R Penrose

12.2.7 Black holes versus white holes

General relativity is a time-symmetric theory. So, to any solution of its
equations (with time-symmetric equations of state) that is asymmetric in
time, there must correspond another solution for which the time-ordering
is reversed.† One of the most familiar solutions is that representing
(spherically symmetric) collapse of a star (described using, say, the T_{ab} of
'dust') to become a black hole.[54,55,7] In time-reversed form this represents
what is referred to as a 'white hole' finally exploding into a cloud of
matter. Spacetime diagrams for the two situations are given in figure 12.5.

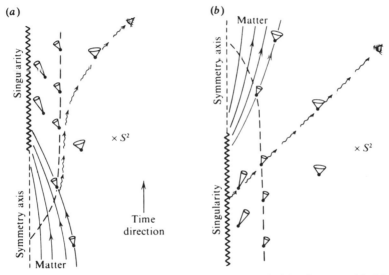

Figure 12.5. Black and white holes (Finkelstein-type picture): (a) collapse to a black hole,
(b) explosion from a white hole. An observer's eye is at the top right.

Various authors[52,53] have attempted to invoke white holes as explanations for quasars or other violent astronomical phenomena (sometimes
using the name of 'lagging cores'). Recall that in a (classical) collapse to a
black hole, the situation starts out with a perfectly normal distribution of
matter which follows deterministic evolutionary equations. At a certain
stage a trapped surface may form, leading to the presence of an *absolute
event horizon* into which particles can fall, but out of which none can
escape. After all the available matter has been swallowed, the hole settles

† For this purpose, a 'solution of the Einstein equations' would be a Lorentzian 4-manifold
with a *time-orientation* (and perhaps a space-orientation). 'Reversing the time-ordering'
amounts to selecting the opposite time-orientation.

down and remains unchanging until the end of time (possibly at recollapse of the universe). (This ignores the quantum-mechanical effects of the Hawking process,[56] which I shall discuss in a moment.) A white hole, therefore, is created at the beginning of time (i.e. in the big bang) and remains in an essentially unchanging state for an indefinite period. Then it disappears by exploding into a cloud of ordinary matter. During the long quiescent period, the boundary of the white hole is a stationary horizon – the *absolute particle horizon* – into which no particle can fall, but out through which particles may, in principle, be ejected.

There is something that seems rather 'thermodynamically unsatisfactory' (or physically improbable) about this supposed behaviour of a white hole, though it is difficult to pin down what seems wrong in a definitive way. The normal picture of collapse to a black hole seems to be 'satisfactory' as regards one's conventional ideas of classical determinism. Assuming that (strong) cosmic censorship[57-59] holds true, the entire spacetime is determined to the *future* of some 'reasonable' Cauchy hypersurface, on which curvatures are small. But in the case of the white hole, there is no way of specifying such boundary conditions in the past because an initial Cauchy hypersurface has to encounter (or get very close to) the singularity. Put another way, the future behaviour of such a white hole does not, in any *sensible* way, seem to be determined by its past. In particular, the precise moment at which the white hole explodes into ordinary matter seems to be entirely of its own 'choosing', being unpredictable by the use of normal physical laws. Of course one could use future boundary conditions to retrodict the white hole's behaviour, but this (in our entropy-increasing universe) is the thermodynamically unnatural way around. (And, in any case, one can resort to memory as a more effective means of retrodiction!)

Related to this is the fact that while an external observer (using normal retarded light) cannot directly see the singularity in the case of a black hole, he can do so in the case of a white hole (figure 12.5). Since a spacetime singularity is supposed to be a place where the known physical laws break down it is perhaps not surprising, then, that this implies a strong element of indeterminism for the white hole. Causal effects of the singularity can, in this case, influence the outside world.

The presently accepted picture of the physical effects that are expected to accompany a spacetime singularity, is that *particle creation* should take place.[62,63,66] This is the general conclusion of various different investigations into curved-space quantum field theory. However, owing to the incomplete state of this theory, these investigations do not always agree on the details of their conclusions. As applied to a white hole, two particular schools of thought have arisen. According to Zel'dovich[64] the white hole ought to be completely unstable to this process, evaporating away instantaneously, while Hawking[67] has put forward the ingenious viewpoint that the white hole evaporation ought to be much lower, and indistinguishable in nature from that produced, according to the Hawking process, by a black hole of the same mass – indeed, that a white hole ought itself to be indistinguishable from a black hole! This Hawking viewpoint is, in a number of respects, a very radical one which carries with it some serious difficulties. I shall consider these in a moment. The other viewpoint has the implication that white holes should not physically exist (though, owing to the tentative nature of the particle-creation calculations, this may not carry a great deal of weight; however, cf. also Eardley[65]).

There is another reason for thinking of white holes as antithermodynamic objects (though this reason, too, must be modified if one attempts to adopt the above-mentioned more radical of Hawking's viewpoints). According to the Bekenstein–Hawking formula,[56,69] the surface area A of a *black* hole's absolute event horizon is proportional to the intrinsic entropy S of the hole:

$$S = kAc^3/4\hbar G$$

(k being Boltzmann's constant and G being Newton's gravitational constant). The area principle of classical general relativity[7,70,71] tells us that A is non-decreasing with time in classical processes, and this is compatible with the thermodynamic time's arrow that entropy should be non-decreasing. Now, if a white hole is likewise to be attributed an intrinsic entropy, it is hard to see how the value of this entropy can be other than that given by the Bekenstein–Hawking formula again, but where A now refers to the absolute *particle* horizon. The time-reverse of the area principle then tells us that A is *non-increasing* in classical processes, which is the opposite of the normal thermodynamic time's arrow for entropy. In particular, the value of A will substantially *decrease* whenever the white hole ejects a substantial amount of matter, such as in the final explosion shown in figure 12.5. Thus, this is a strongly antithermodynamic behaviour.

It seems that there are two main possibilities to be considered concerning the physics of white holes. One of these is that there is a general principle that rules out their existence (or, at least, that renders them overwhelmingly improbable). The other possibility is contained in the aforementioned line of argument due to Hawking, which suggests that because of quantum-mechanical effects, black and white holes are to be regarded as physically indistinguishable.[67] I shall discuss, first, Hawking's remarkable idea. Then I shall attempt to indicate why I nevertheless believe that this cannot be the true explanation, and that it is necessary that white holes do *not* physically exist.

Recall, first, the Hawking radiation that is calculated to accompany any black hole. The temperature of the radiation is inversely proportional to the mass of the hole, being of the general order of 10^{-7} K in the case of a black hole of 1 $M_\odot$. Of course this temperature is utterly insignificant for stellar-mass holes, but it could be relevant observationally for very tiny holes, if such exist. In an otherwise empty universe, the Hawking radiation would cause the black hole to lose mass, become hotter, radiate more, lose more mass, etc., the whole process accelerating until the hole disappears (presumably) in a final explosion. But for black hole of solar mass (or more) the process would take $>10^{53}$ Hubble times! And, so that the process could even begin, a wait of 10^7, or so, Hubble times would be needed to enable the expansion of the universe to reduce the present background radiation to below that of the hole – assuming an indefinitely expanding universe-model!

The absurdity of such figures notwithstanding, it is of some considerable theoretical interest to contemplate, as Hawking has done,[67,72] the state of thermal equilibrium that would be achieved by a black hole in a large container with perfectly reflecting walls. If the container is sufficiently large for a given total mass–energy content (case (*a*)), the black hole will radiate itself away completely (presumably) – after having swallowed whatever other stray matter there had been in the container – to leave, finally, nothing but thermal radiation (with perhaps a few thermalized particles). This final state will be the 'thermal equilibrium' state of maximum entropy (see figure 12.6(*a*)).

If the container is substantially smaller (case (*c*)) – or, alternatively, if the mass–energy content is substantially larger (though still not large enough to collapse the whole container) – the maximum-entropy state will be achieved by a single spherical black hole in thermal equilibrium with its surrounding radiation. Stability is here achieved because, if by a fluctuation the hole radiates a bit too much and consequently heats up, its surroundings heat up even more and cause it to absorb more than it emits and thus to return to its original size; if by a fluctuation it radiates less than it absorbs, its surroundings cool by more than it does and again it returns to equilibrium (figure 12.6(*c*)).

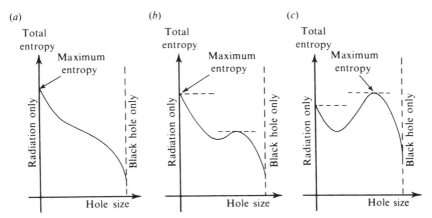

Figure 12.6. Hawking's black hole in a perfectly reflecting container: (*a*) large container, (*b*) intermediate container, (*c*) smallish container.

There is also a situation (case (*b*)) in which the container lies in an intermediate size range, for the given mass–energy content, and where the black hole is still stable, but only represents a *local* entropy maximum, the absolute maximum being a state in which there is only thermal radiation (and perhaps thermalized particles) but no black hole. In this case the black hole can remain in equilibrium with its surrounding radiation for a very long period of time. A large fluctuation would be needed in which a considerable amount of radiation is emitted by the hole, sufficient to get across the low entropy barrier between the two local maxima (figure 12.6(*b*)). With such a large mass loss from the hole, it is able to heat up by an amount greater than its surroundings are able to do; it then loses more mass, heats up more, and, as in case (*a*), radiates itself away completely, to give the required state of thermal radiation (plus occasional thermalized particles).

It should be pointed out that whereas we are dealing with processes that have absurdly long timescales,† these situations have a rather fundamental significance for physics. We are concerned, in fact, with the states of maximum entropy for *all* physical processes. In cases (*a*) and (*b*), the maximum-entropy state is the familiar 'heat death of the universe', but in case (*c*) we have something new: a black hole in thermal equilibrium with radiation. There are, of course, many detailed theoretical difficulties with this setup (e.g. the Brownian motion of the black hole would occasionally send it up against a wall of the container, whereupon

† If we allow (virtual) black holes of down to 10^{-33} cm (i.e. $\sim 10^{-20}$ of an elementary particle radius) then we obtain a picture[73] for which these timescales can be very short.

the container should be destroyed). Such problems will be ignored as irrelevant to the main issues! But also, since the relaxation times are so much greater than the present age of the universe, the interpretations of one's conclusions do need some care. Nevertheless, I feel that there are important insights to be gained here.

To proceed further with Hawking's argument, consider case (*c*). For most of the time the situation remains close to maximum entropy: a black hole with radiation. But occasionally, via an initial large fluctuation in which a considerable energy is emitted by the hole, a sequence like that just considered for case (*b*) will occur, where the black hole evaporates away to give thermal radiation. But then, after a further long wait, enough radiation (again by a fluctuation) collects together in a sufficiently small region for a black hole to form. Provided this hole is large enough, the system settles back into the maximum-entropy state again, where it remains for a very long while.

Cycles like this can also occur in case (*b*) (and even in case (*a*)), but with the difference that most of the time is spent in a state where there is no black hole. In case (*c*), a black hole is present most of the time. Hawking now argues that since the essential physical theories involved are time-symmetric (general relativity, Maxwell theory, neutrinos, possibly electrons, pions, etc., and the general framework of quantum mechanics), the equilibrium states ought to be time-symmetric also. But reversing the time-sense leads to white holes, not to black holes. Thus, Hawking proposes, white holes ought to be physically indistinguishable from black holes!

This identification is not so absurd as one might think at first. The Hawking radiation from the black hole becomes reinterpreted as particle creation near the singularity of the white hole (and hence Hawking proposes a rather slow rate of particle production at the white hole singularity). The swallowing of radiation by the black hole becomes time-reversed Hawking radiation from the white hole. One can, of course, envisage a black hole swallowing a complicated object such as a television set. How can this be thought of as time-reversed Hawking radiation? The argument is that Hawking radiation, being thermal,[67,74] produces all possible configurations with equal probability. It is *possible* to produce a television set as part of the Hawking radiation of a black hole, but such an occurrence is overwhelmingly improbable and would correspond to a large reduction in entropy. A black hole swallowing a television set only seems more 'natural' because we are used to situations in which the entropy is low in the initial state. We can equally well envisage initial boundary conditions of low entropy for the time-reversed Hawking radiation – and this would be the case for a television set being annihilated as time-reversed Hawking radiation of a white hole.

So far, this all seems quite plausible, and there is even a certain unexpected elegance and economy in the whole scheme. But unfortunately is suffers from two (or perhaps three) very severe drawbacks which, in my opinion, rule it out as a serious possibility.

In the first place, whereas the geometry of the spacetime outside a stationary black hole's horizon is identical to that outside a stationary white hole's horizon, it is definitely *not* so that the exterior geometry of a black hole that forms by standard gravitational collapse and then finally disappears according to the Hawking process is time-symmetric. This time-asymmetry is made particularly apparent by use of conformal diagrams as shown in figure 12.7. A precise distinction between the

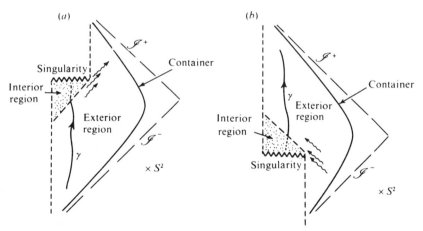

Figure 12.7. Conformal diagrams illustrating time-asymmetry of a transient black hole. (*a*) Classical collapse to a black hole followed by complete Hawking evaporation. (*b*) Hawking condensation to a white hole followed by its classical disappearance.

transient black hole and white hole external geometries can be made in terms of their TIP and TIF structures (cf. section 12.3.2). But intuitively, the distinction should be clear from the presence of timelike curves γ which, in the case of the black hole, 'leave' the exterior geometry to 'enter' the hole; and the other way around in the case of the white hole. The reason for this distinction is simply that the process of *classical* collapse is not the time-reverse of the *quantum* Hawking process. This should not really surprise us since each relies on quite different physical theories (classical general relativity as opposed to quantum field theory on a fixed curved-space background).

A point of view adopted by Hawking which might avoid this difficulty is to regard the spacetime geometry as being somewhat observer-dependent. Thus, as soon as quantum mechanics and curved-space geometry have become essentially intertwined, so this viewpoint would maintain, one cannot consistently talk about a classically objective spacetime manifold. An observer who falls into a white hole to be evaporated away as its time-reversed Hawking radiation would, accordingly, believe the geometry to be, instead, that of a black hole whose horizon he crosses and inside which he awaits his 'classical' fate of final destruction by excessive tidal forces.

I have to say that I find this picture almost as hard to accept as those according to which the entropy starts decreasing when the observer's universe collapses about him. If one is considering black or white holes whose radius is of the order of the Planck length ($\sim 10^{-33}$ cm) – or even, possibly, of the order of an elementary particle size ($\sim 10^{-13}$ cm) – then such indeterminacy in the geometry might be acceptable. But for a black hole of solar mass (or more) this would entail a very radical change in our views about geometry, a change which would drastically affect almost any other application of general relativity to astrophysical phenomena. It is true that in section 12.2.5 I have briefly entertained the possibility of a world-view which allows for an element of 'observer-dependence' in the geometry. But I have as yet seen no way to relate such a view to the kind of indeterminacy in classical geometry that the physical identification of black holes with white holes seems to lead to.

But there are also other objections to attempting to regard classical gravitational collapse as being effectively the time-reverse of a quantum-mechanical particle creation process. One of these refers not so much to the attempted identification of the Hawking process with the time-reverse of the classical swallowing of matter by a black hole, but with the attempt to identify either of these processes with the phenomenon of particle production at regions of large spacetime curvature. Such a further identification *seems* to be an integral part of the time-symmetric view that I have been discussing, though it may well not really be what is intended. I am referring to the picture of Hawking radiation by a black hole as being alternatively regarded as a process of particle production near the singularity of a white hole. If, indeed, it can be so regarded, then this is not the 'normal' process of particle production at regions of large curvature that has many times been discussed in the literature.[62] For in

As Statues moulder into Worth. *P. II.*

To Nature and your Self appeal,
Nor learn of others, what to feel.

Anon:

VIII. *The relativity of aging.* Time Smoking a Picture, by William Hogarth (1761). This is a satire on the idea that old things are more valuable than new ones. Hogarth likens Time to an unreliable picture dealer who blows smoke on the landscape and drives his scythe through it.

that process, particles are always produced in *pairs*: baryon with anti-baryon, lepton with antilepton; positively charged particle with negatively charged particle. But the Hawking process is explicitly not of this form, as its thermal nature (for particles escaping to infinity) implies.[67,74,75]

The contrast is even more blatant if we try to relate this pairwise particle production near white hole singularities to the time-reverse of the destruction of matter at a black hole's singularity. For there are no constraints whatever on the type of matter that a black hole can classically absorb. And if strong cosmic censorship is accepted in classical processes, it seems that even individual charged particles have then to be separately destroyed at the singularity. (This point will be amplified in section 12.3.2.) There is no suggestion that the particles must somehow contrive to sort themselves into particle–antiparticle pairs before they encounter the singularity. The difficulty here is, perhaps, not so much directly to do with Hawking's identification of black holes with white holes, but with the whole idea of hoping to deal with the matter destruction–creation process in terms of known physics. Thus I maintain that, whereas it *may* be that matter creation at the big bang can be treated in terms of known (or at least partially understood) particle-creation processes, this seems *not* to be true of the destruction processes at black hole singularities – nor, if Hawking is right, of the creation processes at white hole singularities. Thus, the Hawking view would seem to lead to direct conflict with the often expressed hope that particle creation at the big bang can be understood in terms of processes of particle production by spacetime curvature. This relates to the question of whether time-symmetric physics can be maintained at spacetime singularities. This is a key issue that I shall discuss in more detail in section 12.3.

More directly related to Hawking's proposal is a difficulty which arises if we examine in detail the cycles whereby a solar-mass black hole (say), in stable equilibrium with its surroundings, in our perfectly reflecting container, may disappear and reform, owing to fluctuations, in cases (*b*) and (*c*) just discussed. What, in fact, is the most probable way for the black hole to evaporate completely? It might, of course, simply throw out its entire mass in one gigantic fluctuation. But the random Hawking process would achieve this only absurdly infrequently. Overwhelmingly *less* absurdly infrequent would be the emission, in one huge fluctuation, of that *fraction* of the hole's mass needed to raise its Bekenstein–Hawking temperature above that to which its surroundings would be consequently raised. From there on, the evaporation would proceed

167

'normally' needing no further improbable occurrence. But consider the final explosion according to which the hole 'normally' disappears. Electrons and positrons have been appearing, followed by pions; then, at the last moment, a whole host of unstable particles is produced which undergo complicated decays. Finally one expects many protons and antiprotons separately escaping the annihilation point. Only much later, by chance encounters as they move randomly about the container, would one expect the protons and antiprotons gradually to annihilate one another (or possibly occasionally to decay, themselves, by a Pati–Salam-type process[76] into, say, positrons and into electrons which would then mostly annihilate one another).

What, now, must we regard as the most probable way in which a black hole forms again in the container, to reach another point of stable equilibrium with its surroundings? Surely it is *not* the time-reverse of the above, according to which the protons and antiprotons must first (with long preparation) contrive to form themselves (quite unnecessarily) out of the background radiation before aiming themselves with immense accuracy at a tiny point, only to indulge in some (again unnecessary) highly contrived particle physics whereby they meet up with carefully aimed γ-rays, etc., to form various unstable particles, etc., etc.; and then (also with choreography of the utmost precision) other particles (pions, then electrons and positrons) must aim themselves inwards, having first formed themselves out of the background at the right moment and in the correct proportions. Only later does the background radiation itself fall inwards to form the bulk of the mass needed to form the hole.

My point is not that this curious beginning is necessarily the most improbable part of the process. I can imagine that it may well not be. But it is unnecessary. The essential part of the hole formation of a *black* hole would occur when the radiation itself collects together in a sufficiently small region to undergo what is, in effect, a standard gravitational collapse. In fact, it would seem that the tiny core that has been formed with such elaborate preparations ought somewhat to *inhibit* the subsequent inward collapse, owing to its excessive temperature!

So what has gone wrong? What time-asymmetric physics has been smuggled into the description of the 'most probable' mode of disappearance of the hole, that it should so disagree with the time-reverse of its 'most probable' mode of reappearance? Possibly none, if white holes, in principle, exist, and are simply *different* objects from black holes. While the above elaborate preparations are not necessary for the formation of a black hole, they could be for a white hole. After all, one has to contrive some way of producing the white hole's initial singularity which, as is evident from figure 12.7, has a quite different structure from that of a black hole. It might be that the production of such a singularity is an extraordinarily delicate process, requiring particles of just the right kind and in an essentially right order to be aimed with high energy and with extraordinary accuracy. There is an additional seeming difficulty, however, in that one also has to conjure up a region of spacetime (namely that inside the horizon) which does not lie in the domain of dependence of some initial Cauchy hypersurface drawn before the white hole appears (figure 12.7(*b*)). Of course, the time-reverse of this problem also occurs for the (Hawking) disappearance of a black hole, but one is not in the habit of trying to retrodict from Cauchy hypersurfaces, so this seems less worrisome! An additional difficulty, if white holes are allowed, is that we now encounter the problem, mentioned earlier, of trying to predict what the white hole is going to emit, and when. As I have indicated, *if* time-symmetric physics holds at singularities, then 'normal' ideas of particle creation due to curvature will not do. Hawking's concept of 'randomicity'[67] might be nearer the mark, but it is unfortunately too vague to enable any calculations to be made – now that the essential guiding idea of an identification between black holes and white holes has been removed.

I find the picture of an equilibrium involving the occasional production of such white holes a very unpleasant one. And what other monstrous zebroid combinations of black and white might also have to be contemplated? I feel that such things have nothing really to do with physics (at least on the macroscopic scale). The only reason that we have had to consider white holes *at all* is in order to save time-symmetry! The consequent unpleasantness and unpredictability seems a high price to pay for something that is *not even true* of our universe on a large scale.

One of the consequences of the hypothesis that I shall set forth in the next section is that it rules out the white hole's singularity as an unacceptable boundary condition. The hypothesis is time-asymmetric, but this is necessary in order to explain the other arrows of time. When we add this hypothesis to the discussion of equilibrium within the perfectly reflecting container, we see explicitly what time-asymmetric physics has been 'smuggled' in. For the hypothesis is designed not to constrain the behaviour of black holes in any way, but it forbids white holes and therefore renders irrelevant the extraordinary scenario that we seem to need in order to produce one!

I hope the reader will forgive me for having discussed white holes at such length only to end by claiming that they do not exist! But hypothetical situations can often lead to important understandings, especially when they border on the paradoxical, as seems to be the case here.

12.3 Singularities: the key?

What is the upshot of the discussion so far? According to sections 12.2.3, 12.2.4 and 12.2.5, the arrows of entropy and retarded radiation, and *possibly* of psychological time, can all be explained if a reason is found for the initial state of the universe (big bang singularity) to be of comparatively low entropy and for the final state to be of high entropy. According to section 12.2.6, some low-entropy assumption *does* need to be *imposed* on the big bang; that is, the mere fact that the universe expands away from a singularity is in no way sufficient. And according to section 12.2.7, we need some assumption on initial singularities that rules out those which would lie at the centres of white holes. On the other hand, the discussions in sections 12.2.1 and 12.2.2 were inconclusive, and I shall need to return to them briefly at the end.

But what is it in the nature of the big bang that is of 'low entropy'? At first sight, it would seem that the knowledge we have of the big bang points in the opposite direction. The matter (including radiation) in the early stages appears to have been completely thermalized (at least so far as this is possible, compatibly with the expansion). If this had not been so, one would not get correct answers for the helium abundance, etc.[77,78] And it is often remarked upon that the 'entropy per baryon' (i.e. the ratio of photons to baryons) in the universe has the 'high' value of $\sim 10^9$. Ignoring the contribution to the entropy due to black holes, this value has remained roughly constant since the very early stages, and then represents easily the major contribution to the entropy of the universe – despite all the 'interesting' processes going on in the world, so important to our life here on Earth, that depend upon 'small' further taking up of entropy by stars like our Sun. The answer to this apparent paradox – that the big bang thus *seems* to represent a state of *high* entropy – lies in the unusual nature of gravitational entropy. This I next discuss, and then show how this relates to the structure of singularities.

12.3.1 Gravitational entropy

It has been pointed out by many authors[79] that gravity behaves in a somewhat anomalous way with regard to entropy. This is true just as much for Newtonian theory as for general relativity. (In fact, the situation is rather worse for Newtonian theory.) Thus, in many circumstances in which gravity is involved, a system may behave as though it has a negative specific heat. This is directly true in the case of a black hole emitting Hawking radiation, since the more it emits, the hotter it gets. But even in such familiar situations as a satellite in orbit about the Earth, we observe a phenomenon of this kind. For dissipation (in the form of frictional effects in the atmosphere) will cause the satellite to speed up, rather than slow down, i.e. cause the kinetic energy to increase.

This is essentially an effect of the universally attractive nature of the gravitational interaction. As a gravitating system 'relaxes' more and more, velocities increase and the sources clump together – instead of uniformly spreading throughout space in a more familiar high-entropy arrangement. With other types of force, their attractive aspects tend to saturate (such as with a system bound electromagnetically), but this is not the case with gravity. Only non-gravitational forces can prevent parts of a gravitationally bound system from collapsing further inwards as the system relaxes. Kinetic energy itself can halt collapse only temporarily. In the absence of significant non-gravitational forces, when dissipative effects come further into play, clumping becomes more and more marked as the entropy increases. Finally, maximum entropy is achieved with collapse to a black hole – and this leads us back into the discussion of section 12.2.7.

Consider a universe that expands from a 'big bang' singularity and then recollapses to an all-embracing final singularity. As was argued in section 12.2.6, the entropy in the late stages ought to be much higher than the entropy in the early stages. How does this increase in entropy manifest itself? In what way does the high entropy of the final singularity distinguish it from the big bang, with its comparatively low entropy? We may suppose that, as is apparently the case with the actual universe, the entropy in the initial *matter* is high. The kinetic energy of the big bang, also, is easily sufficient (at least on average) to overcome the attraction due to gravity, and the universe expands. But then, relentlessly, gravity begins to win out. The precise moment at which it does so, locally, depends upon the degree of irregularity already present, and probably on various other unknown factors. Then clumping occurs, resulting in clusters of galaxies, galaxies themselves, globular clusters, ordinary stars, planets, white dwarfs, neutron stars, black holes, etc. The elaborate and interesting structures that we are familiar with all owe their existence to this clumping, whereby the gravitational potential energy begins to be taken up and the entropy can consequently begin to rise above the *apparently* very high value that the system had initially. This clumping must be expected to increase; more black holes are formed; smallish black holes swallow material and congeal with each other to form bigger ones. This process accelerates in the final stages of recollapse when the average density becomes very large again, and one must expect a very irregular and clumpy final state.

There is a slight technical difficulty in that the concept of a black hole is normally only defined for asymptotically flat (or otherwise open) spacetimes. This difficulty could affect the discussion of the final stages of collapse when black holes begin to congeal with one another, and with the final all-embracing singularity of recollapse. But I am not really concerned with the location of the black holes' event horizons, and it is only in precisely defining these that the aforementioned difficulty arises. A black hole that is formed early in the universe's history has a singularity that is reached at early proper times for observers who encounter it;[57] for holes that are formed later, they can be reached at later proper times. On the basis of strong cosmic censorship (cf. section 12.3.2), one expects all these singularities eventually to link up with the final singularity of recollapse.[57] I do not require that the singularities of black holes be, in any clear-cut way, distinguishable from each other or from the final singularity of recollapse. The point is merely that the gravitational clumping which is characteristic of a state of high gravitational entropy should manifest itself in a very complicated structure for the final singularity (or singularities).

The picture is not altogether dissimilar for a universe that continues to expand indefinitely away from its big bang. We still expect local clumping, and (provided that the initial density is not altogether too low or too uniform for galaxies to form at all) a certain number of black holes should arise. For the regions inside these black holes, the situation is not essentially different from that inside a collapsing universe (as was remarked upon in section 12.2.6), so we expect to find, inside each hole, a very complicated singularity corresponding to a very high gravitational entropy. For those regions not inside black holes there will still be certain localized portions, such as rocks, planets, black dwarfs, or neutron stars, which represent a certain ultimate raising of the entropy level owing to gravitational clumping, but the gain in gravitational entropy will be relatively modest, though sufficient, apparently, for all that we need for life here on Earth.

I have been emphasizing a qualitative relation between gravitational clumping and an entropy increase due to the taking up of gravitational potential energy. In terms of spacetime curvature, the absence of clumping corresponds, very roughly, to the absence of Weyl conformal curvature (since absence of clumping implies spatial-isotropy, and hence no gravitational principal null-directions).[45] When clumping takes place, each clump is surrounded by a region of nonzero Weyl curvature. As the clumping gets more pronounced owing to gravitational contraction, new regions of empty space appear with Weyl curvature of greatly increased magnitude. Finally, when gravitational collapse takes place and a black hole forms, the Weyl curvature in the interior region is larger still and diverges to infinity at the singularity.

At least, that is the picture presented in spherically symmetric collapse, the magnitude of the Weyl curvature diverging as the inverse cube of the distance from the centre. But there are various reasons for believing that in generic collapse, also, the Weyl curvature should diverge to infinity at the singularity, and (at most places near the singularity) should dominate completely over the Ricci curvature.

This can be seen explicitly in the details of the Belinskii–Khalatnikov–Lifshitz analysis.[80] Moreover, one can also infer it on crude qualitative grounds. In the exact Friedmann models, it is true, the Ricci tensor dominates, the Weyl tensor being zero throughout. In these cases, as a matter world-line is followed into the singularity, it is approached isotropically by the neighbouring matter world-lines, so we have simultaneous convergence in three mutually perpendicular directions orthogonal to the world-line. In the case of spherically symmetrical collapse to a black hole, on the other hand, if we envisage some further matter falling symmetrically into the central singularity, it will normally converge in towards a given matter world-line only in *two* mutually perpendicular directions orthogonal to the world-line (and diverge in the third). This is the situation of the Kantowski–Sachs[81] cosmological model, giving a so-called 'cigar'-type singularity.[7] If r is the usual Schwarzschild coordinate, the volume gets reduced like $r^{3/2}$ near the singularity, so the densities are $\sim r^{-3/2}$. Thus, for a typical Ricci tensor component, $\Phi \sim r^{-3/2}$. However, in general, for a typical Weyl tensor component, $\Psi \sim r^{-3}$, showing that the Weyl tensor dominates near the singularity in these situations. Also, in the 'pancake' type of singularity, where there is convergence in only *one* direction orthogonal to a matter world-line, we again expect the Weyl tensor to dominate with $\Phi \sim r^{-1}$ and $\Psi \sim r^{-2}$ in this case.

173

Now the Friedmann type of situation, with simultaneous convergence of all matter from all directions at once, would seem to be a very special setup. If there is somewhat less convergence in one direction than in the other two, then a cigar-type configuration seems more probable very close to the singularity, while a pancake-type appears to result when the main convergence is only in one direction. Moreover, with a generic setup, a considerable amount of oscillation seems probable.[80] An oscillating Weyl curvature of frequency ν and complex amplitude Ψ, supplies an *effective* additional 'gravitational-energy' contribution to the Ricci tensor[61] of magnitude $\sim |\Psi|^2 \nu^{-2}$. If ν becomes very large so that many oscillations occur before the singularity is reached, then[49] $\nu^2 \gg \Phi^{-1}$, where Φ is a typical Ricci tensor component. Thus if, as seems reasonable in general, the 'energy content' of Ψ is to be comparable with Φ as the singularity is approached, we have $|\Psi|^2 \nu^{-2} \sim \Phi$, so $|\Psi| \gg \Phi$. These considerations are very rough, it is true, but they seem to concur with more detailed analysis[80] which indicates that in generic behaviour near singularities the contributions due to matter can be ignored to a first approximation and the solution treated as though it were a vacuum, i.e. that the Weyl part of the curvature dominates over the Ricci part.

The indications are, then, that a high-entropy singularity should involve a very large Weyl curvature, unlike the situation of the singularity in the Friedmann dust-filled universe or any other models of the Robertson–Walker class. At the time of writing, however, no clear-cut integral formula (say) which could be regarded as giving mathematical expression to this suggested relation between Weyl curvature and gravitational entropy has come to light. Some clues as to the nature of such a formula (if such exists at all) may be obtained, firstly, from the Bekenstein–Hawking formula for the entropy of a black hole and, secondly, from the expression for the particle number operator for a linear spin-2 massless quantized free field – since an estimate of the 'number of gravitons' in a gravitational field could be taken as a measure of its entropy.† Thus this entropy measures the number of *quantum* states that contribute to a given classical geometry.

There is one final point that should be mentioned in connection with the question of the entropy in the gravitational field. It was pointed out some time ago by Tolman[84] that a model universe containing matter that

---

† This point of view does not seem to agree with that of Gibbons and Hawking,[82] who apparently regard the gravitational entropy as being zero when black holes are absent. But an estimate of 'photon number' in a classical electromagnetic field gives a measure of its entropy[83] (without black holes). Gravity is presumably similar.

appeared to be in thermal equilibrium in its early stages can lead to a situation in which the matter gets out of equilibrium as the universe expands (a specific example of matter illustrating such behaviour being a diatomic gas which is capable of dissociating into its elements and recombining). Then, if such a model represents an expanding and recollapsing universe, the state of the matter during recollapse would differ from the corresponding state during expansion, where we make the correspondence at equal values of the universe radius R (or comoving radius R). In fact, the matter, during recollapse, would have acquired some energy out of the global geometry of the universe, the resulting difference in geometry showing up in the fact that $\dot{R}^2$ is greater, for given R, at recollapse than it is during the expansion. So the entropy of the system as a *whole* increases with time even though the matter *itself* is in thermal equilibrium during an initial stage of the expansion. There is, in fact, a contribution to the entropy from R (and $\dot{R}$), which must be regarded as a dynamical variable in the model. (This arises because of the phenomenon of *bulk viscosity*.[85])

One can view what is involved here as basically a transfer of potential energy from the global structure of the universe (gravitational potential energy) into the local energy of the matter, though there are well-known difficulties about defining energy in a precise way for models of this kind. But these difficulties should not concern us unduly here, since it is actually the entropy rather than the energy that is really relevant, and entropy has much more to do with probabilities and coarse-graining than it has to do with any particular definition of energy. In the example given by Tolman there is no state of maximum entropy, either achieved in any one specific model or throughout all models of this type. By choosing the value of R at maximum expansion to be sufficiently large (for fixed matter content), the total entropy can be made as large as we please. Tolman envisaged successive cycles of an 'oscillating' universe with gradually increasing maximum values of R. However, it is hard for us to maintain such a world-view now, because the singularity theorems[7,8] tell us that the universe cannot achieve an effective 'bounce' at minimum radius without violating the known† laws of physics.

From my own point of view, the situation envisaged by Tolman may be regarded as one aspect of the question of how the structure of the universe as a whole contributes to the entropy. It apparently concerns a somewhat different aspect of this question than does gravitational

† I am counting quantum gravity as 'unknown' whether or not it helps with the singularity problem!

clumping, since the Weyl tensor is everywhere zero in Tolman's models. It is clear that this has also to be understood in detail if we are to perceive, fully, the role of gravitational entropy. Nevertheless, it appears that the entropy available in Tolman's type of situation is relatively insignificant[78] compared with that which can be obtained – and, indeed, *is* obtained – by gravitational clumping (cf. section 12.3.3).

The key question must ultimately concern the structure of the singularities. These singularities, in any case, provide the boundary conditions for the various cycles in Tolman's 'oscillating' universe. Moreover, as we shall see in a moment, if strong cosmic censorship holds true, the presence of irregularities should not alter the all-embracing nature of these singularities in the case of an expanding and recollapsing universe.

(7) Hawking, S. W. and Ellis, G. F. R. (1973). *The Large Scale Structure of Space–time.* Cambridge University Press.

(8) Hawking, S. W. and Penrose, R. (1970). *Proc. R. Soc. Lond.,* **A314**, 529.

(14) Aharanov, Y., Bergmann, P. G. and Lebowitz, J. L. (1964). *Phys. Rev.,* **B134**, 1410.

(45) Penrose, R. (1964). In *Relativity, Groups and Topology,* eds. C. M. DeWitt and B. S. DeWitt. Gordon & Breach: New York; and in Gold, (1967) [cited on p. 137 of this book].

(52) Novikov, I. D. (1964). *Astron. Zh.,* **41**. (English translation in *Sov. Astron.: Astron. J.,* **8**, 857, 1075.)

(53) Ne'eman, Y. (1965). *Astrophys. J.,* **141**, 1303.

(54) Oppenheimer, J. R. and Synder, H. (1939). *Phys. Rev.,* **56**, 455.

(55) Misner, C. W., Thorne, K. S. and Wheeler, J. A. (1973). *Gravitation.* Freeman: San Francisco.

(56) Hawking, S. W. (1975). *Commun. Math. Phys.,* **43**, 199.

(57) Penrose, R. (1974). In *Confrontation of Cosmological Theories with Observational Data.* (IAU Symp. 63), ed. M. S. Longair. Reidel: Boston.

(58) Penrose, R. (1978). In *Theoretical Principles in Astrophysics and Relativity,* eds. N. R. Lebovitz, W. H. Reid and P. O. Vandervoort. University of Chicago Press.

(59) Penrose, R. (1977). In *Proceedings of the First Marcel Grossmann Meeting on General Relativity, ICTP Trieste,* ed. R. Ruffini, North-Holland: Amsterdam.

(61) Penrose, R. (1966). In *Perspectives in Geometry and Relativity,* ed. B. Hoffmann. Indiana University Press.

(62) Parker, L. (1977). In *Asymptotic Structure of Space–Time,* eds. F. P. Esposito and L. Witten. Plenum: New York.

(63) Sexl, R. U. and Urbantke, H. K. (1969). *Phys. Rev.,* **179**, 1247.

(64) Zel'dovich, Ya. B. (1974). In *Gravitational Radiation and Gravitational Collapse* (IAU Symp. 64), ed. C. M. DeWitt. Reidel: Boston.

(65) Eardley, D. M. (1974). *Phys. Rev. Lett.,* **33**, 442.

References

(66) Zel'dovich, Ya. B. and A. A. Starobinsky (1971). *Zh. Eksp. Teor. Fiz.*, **61**, 2161. (English translation in *Sov. Phys.: JETP*, **34**, 1159.)

(67) Hawking, S. W. (1976). *Phys. Rev.*, **D13**, 191; **14**, 2460.

(69) Bekenstein, J. D. (1973). *Phys. Rev.*, **D7**, 2333; (1974) **9**, 3292.

(72) Gibbons, G. W. and Perry, M. J. (1978). *Proc. R. Soc. Lond.*, **A358**, 467.

(73) Wheeler, J.A. (1962). *Geometrodynamics*. Academic Press: London, New York.

(74) Wald, R. M. (1975). *Commun. Math. Phys.*, **45**, 9.

(75) Parker, L. (1975). *Phys. Rev.*, **D12**, 1519.

(76) Pati, J. C. and Salam, A. (1973). *Phys. Rev. Lett.*, **31**, 661; *Phys. Rev.*, **D8**, 1240; (1974) **10**, 275.

(78) Weinberg, S. (1972). *Gravitation and Cosmology*. Wiley: New York.

(79) Lynden-Bell, D. and Lynden-Bell, R. M. (1977). *Mon. Not. R. Astron. Soc.*, **181**, 405.

(80) Belinskii, V. A., Khalatnikov, I. M. and Lifshitz, E. M. (1970). *Adv. Phys.*, **19**, 525.

(81) Kantowski, R. and Sachs, R. K. (1967). *J. Math. Phys.*, **7**, 443.

(82) Gibbons, G. W. and Hawking, S. W. (1977). *Phys. Rev.*, **D15**, 2752.

(83) Zel'dovich, Ya. B. Personal Communication.

(84) Tolman, R. C. (1934). *Relativity, Thermodynamics and Cosmology*. Clarendon Press: Oxford.

(85) Israel, W. (1963). *J. Math. Phys.*, **4**, 1163.

177

BLACK HOLE THERMODYNAMICS AND TIME ASYMMETRY

P. C. W. Davies

Department of Mathematics, University of London King's College,
Strand, London WC2R 2LS

(Received 1976 April 9; in original form 1976 February 9)

SUMMARY

The role of the gravitational field as a source of entropy is discussed, first in connection with cosmology, then for black holes. A review is given of the need for an assumption of ' molecular ' chaos or randomness at the initial cosmological singularity, in order to achieve consistency of statistical mechanics with the observed time asymmetry in the universe. It is argued that a simple randomness assumption cannot always be made, because several singularities may be causally connected. The situation is compared with that of quantum black and white holes confined in a closed box. The possibility of black-hole fluctuations is discussed, together with Hawking's conjecture that black and white holes are indistinguishable.

I. GRAVITY AS A SOURCE OF FREE ENERGY

There is a strange paradox of physical theory which is rediscovered from time to time, concerning the organization of matter and energy in the Universe. In its earliest form, the paradox was simply the question: how did the Universe become so highly ordered? It is well known that thermodynamically the cosmos is far from equilibrium, a fact most conspicuously manifested by the presence of large quantities of high-temperature energy in stars which are surrounded by cold, dark space. The presence of life on Earth is entirely dependent upon the thermodynamic disequilibrium in the vicinity of the Sun which is a consequence of this fact.

A long time ago, Boltzmann proposed that the ordered structure in the Universe arose as a result of a stupendously rare statistical fluctuation of all the atoms and radiation, such as are expected to occur in a closed thermodynamic system according to a theorem of Poincaré. The cycle time for such an extraordinary event is of the ludicrous order $10^{10^{80}}$ yr at least, but our participation in such a staggeringly rare occurrence was explained by Boltzmann as a biological selection effect—life could only form during a period of thermodynamic disequilibrium.

There are many objections to Boltzmann's explanation, not least of which is the fact that most cosmologists prefer to believe in a model of the Universe which features a big-bang ' creation ' event, dated at a mere 10^{10} or so yr ago.

The present macroscopic orderliness of matter and energy could be attributed to the way in which the primordial fireball was set up in the first place, but there is good observational evidence against this. It is generally accepted, on the basis of the existence of (probably) thermal microwave, cosmic background radiation, that in the early stages of the primordial fireball the cosmological material formed a fluid in local thermodynamic equilibrium.

If this is the case, then the paradox of the currently observed organization of the Universe prompts the question: how did cosmological material begin in thermodynamic equilibrium, and then pass in such a short period of time (compared to the Poincaré cycle time) to a condition of disequilibrium, when the second law of thermodynamics would appear to require the opposite?

The resolution of this paradox is by now well known, and has been discussed in some depth in an earlier publication (Davies 1974, hereafter referred to as I). The second law only refers to isolated systems, and there is an important sense in which the Universe does not behave like an isolated system. To appreciate this, first consider an idealized laboratory thermodynamic system consisting of a gas confined in an adiabatic enclosure consisting of a movable piston inside a sealed cylinder. Initially the gas is in equilibrium, but if the piston is *rapidly* moved to increase the volume of the enclosure, the gas will expend in a non-equilibrium fashion and the entropy will increase. If the piston is moved back, to return the gas to its former volume, the internal energy of the gas will have increased. The source of this energy is the external energy supply which is used to move the piston, and in principle the internal energy and the entropy of the gas will continue to rise if the process is repeated cyclically, until the external energy supply is exhausted. In a real gas the actual cause of the entropy increase would be due to many processes, if the relaxation times are more than the piston cycle time (hence the need to expand the gas rapidly). For example, in a multi-component fluid (say non-relativistic matter and radiation) a redistribution of energy between the components will take place, but slowly enough that the gas readjustments always lag somewhat behind the piston motion. It is the time asymmetry of this lag (the gas responds *after* the piston moves) that accounts for the apparently ' irreversible ' increase in entropy of the gas.

It is not sufficient that the gas merely be expanded rapidly to bring about the entropy increase however. It is also necessary to make some assumption about the microscopic condition of the gas. This could be Boltzmann's assumption of molecular chaos, for instance. The important point is that the microscopic motions of the gas particles must not be correlated in any way with the macroscopic motion of the piston.

If the piston is driven by an external source of energy, then the dynamics of the piston will be quite independent of the dynamics of the gas, so that it would be exceedingly unlikely that at the precise moment when the piston moved to expand the gas, if this is chosen *at random*, the gas molecules had conspired to reorganize themselves in anticipation of this motion, rendering the above-mentioned lag a negative one, and so *reducing* the entropy. Expressed differently, we might say that the microscopic condition of the gas is chosen *at random* at the outset of the experiment.

On the other hand, if the gas + piston and cylinder + energy supply is considered in its entirety as an isolated system, then, once the energy supply has failed and the system has come into equilibrium, it would happen on exceedingly rare occasions that a random fluctuation in the molecules of the piston etc. would produce an expansion of the gas with an anticipatory motion of the gas molecules (thereby delivering energy spontaneously to the depleted energy supply at the expense of the gas, the entropy of which is reduced as a consequence). This extraordinary circumstance would occur equally often with the opposite sequence, where a spontaneous motion of the piston would be accompanied by an increase in the entropy of the gas. This equality is expected on the grounds of the overall time symmetry of the isolated total system.

Although there are important differences of detail, the laboratory piston and cylinder mechanism illustrates some important principles concerning entropy in the universe, where the role of the piston is played by the expanding (and perhaps recontracting) space-time and the external energy supply is the cosmological gravitational field. That is to say, the fact that the cosmological material is situated in an expanding Universe has the effect of its behaving like a non-isolated system, with gravity acting as an external source of free energy.

In the early stages of the primordial fireball the cosmological fluid was highly compressed, and the process relaxation times were very short compared to the expansion time scale. Thus, the expansion was essentially quasi-static and isentropic. However, the situation is now reversed. Entropy is being produced because various internal forces of the cosmological fluid are causing rearrangements of matter, and the redistribution of energy among the different constituents. By far the most important such process is the flow of energy from non-relativistic matter to electromagnetic radiation, which is taking placed as a result of nucleosynthesis in stars. Because of the existence of the Coulomb barrier between nuclei, this process is many orders of magnitude slower than the expansion time of the Universe (Hubble time) so that the enormous resulting lag behind equilibrium conditions has produced precisely the disequilibrium already noted, i.e. hot stars radiating into cold space.

The connection between the gravitational field and the entropy density of the cosmological fluid was first investigated by Tolman (1934), who considered within the context of general relativity the effect on the fluid of cyclic expansion and recontraction of the universe, such as has been discussed more recently by a number of cosmologists (Landsberg & Park 1975). Unlike the case of the laboratory piston and cylinder, where the energy source to the piston wiil eventually come into equilibrium at a maximum entropy (the piston being gradually damped to rest when the energy supply fails) the Universe cannot be in equilibrium with its gravitational field, but will always tend to expand or collapse, due to the purely attractive nature of the gravitational force. Tolman therefore found that the entropy of the cosmological fluid in a cyclic universe would continue to increase without limit, and the motion of this Universe, far from being damped out, actually increases in amplitude. We have here the first indication that in some sense the unstable nature of classical gravity leads to an entropy divergence.

2. RANDOMNESS AND THE SINGULARITIES

As in the case of the piston and cylinder, the increase of entropy in the Universe which follows after the expansion requires an assumption about the microscopic condition of the particles which emerge from the singularity or creation event. In I (pp. 197–200) it was proposed that the state which emerges from the initial singularity is a completely random one, in analogy with Boltzmann's condition of molecular chaos for a closed box of gas. This means that the microscopic motions of the cosmological fluid (subject mainly to weak, strong and electromagnetic forces in the early stages) are not correlated with the global dynamics of the Universe moving in accordance with the field equations of general relativity. It follows that, with exceedingly high probability, the entropy-increasing processes will then occur as the fluid lags behind the cosmological motion in its attempt to restore equilibrium.

The meaning of ' randomness ' for the emerging particles is more transparent in the case that a whole ensemble of singularities is available. In I some examples of such a possibility were given. One such example is the case of the cyclic Universe, in which recontraction results in a succession of singularities between cycles. Under these circumstances, the randomness assumption amounts to saying that at the outset of each cycle every conceivable state of the cosmological matter is equally probable. By far the greatest number of such states will result in entropy increasing worlds. Some cycles will actually involve ' anticipatory ' motions of the fluid, resulting in entropy decreasing worlds, while those in which the entropy remains unchanged probably form a set of measure zero among the set of all possible initial states.

Within the context of classical physics, it is thought likely that the initial cosmological singularity (or the many end-point singularities in the cyclic model) are the only possible instances of naked singularities. (Penrose 1974 in a discussion of some of the ideas described here, has suggested a definition of naked singularity which excludes the initial cosmological singularity. That definition is not being used here.) Recently, it has been suggested by Hawking (1975) that quantum effects can lead to the production of naked singularities as the final state of black holes, although it is still far from clear in what way quantum theory will modify the present classical picture of singularities which is being discussed here. If black holes do end in naked singularities, it would be sensible to attempt to extend the assumption of randomness of emerging matter to these instances as well, and indeed, such an assumption has been made by Hawking and dignified with the title ' Randomicity Principle '. The subject of black holes will be dealt with in Sections 3 and 4.

The randomness assumption for the primordial fireball has the effect that physical influences emanating from different directions of the Universe which are reaching the Earth now (say electromagnetic waves) are uncorrelated. That is why we do not observe advanced radiation, for example, which would require coherent incoming wavefronts originating in regions of the Universe which were causally decoupled in the fireball due to the existence of particle horizons. The absence of correlations from different regions of the Universe has been invoked by Penrose & Percival (1962) in a general discussion of time asymmetry, and is referred to as the ' Law of Conditional Independence '.

The idea of a random initial cosmological state has also been proposed by Layzer (1976, and references therein), who describes the microscopic condition of the cosmological matter which emerges from the singularity as a state in which all microscopic information is absent.

Some authors have proposed a direct contradiction to the randomness assumption. For example, Gold (1962) has suggested a model Universe in which matter emerges from the initial singularity with such strong correlations that at a later epoch all the atoms in the Universe continue to start decreasing the entropy. This Universe therefore displays a time symmetry in which a period of entropy increase is followed by a period of entropy decrease, ending symmetrically in a final singularity which is, at least on a macroscopic scale, the time reverse of the initial one.

The situation in which the Universe contains singularities at both temporal extremities poses a problem for the randomness assumption. Clearly it is not possible to choose states of the cosmological fluid independently at both extremities on account of the fact that they are causally connected. If the real, observable Universe recontracts to a singular condition, it will have a final state which is qualitatively sharply distinguishable from the initial state, on account of the higher entropy there. This entropy production, which proceeds throughout the cycle of expansion and contraction, endows the Universe with a global time asymmetry, and leads to a distinction between ' initial ' and ' final ' singularities. Calling these low- and high-entropy singularities, respectively A and B to avoid confusion, we may envisage the behaviour of the Universe in reversed time, with B as the ' initial ' state. We should then be forced to admit that the microscopic state of the cosmological material which emerged from B was a highly special one, containing strong correlations which during the subsequent cycle of expansion and recontraction contrived to reduce the entropy of the Universe by causing heat and light to flow from cold objects into hot stars, etc. In short, the randomness assumption cannot be applied to *both* A and B.

In an idealized Friedmann model, a recontracting universe contains one singularity at each temporal extremity. To each possible singularity A there corresponds a causally connected partner B. Although clearly both A and B cannot be chosen independently at random, the coupled pair (A, B) together may still be picked randomly from some boundary data set. In general, the entropy of the cosmological fluid will be different near A and B, because there are entropy-producing processes taking place during the cosmological expansion and recontraction. There is an equal probability that the entropy of the fluid near A will exceed that near B as vice versa. The set for which the entropy is the same probably is of measure zero in the space of all boundary data. One simply *defines* the ' initial ' singularity as that associated with a lower entropy. In general, *one* of the members of the pair (A, B) will be associated with a highly disordered state.

In a realistic model, the universe may contain singularities with more structure, perhaps with delayed ' bangs ' (white holes), premature collapse (black holes), or even naked singularities formed as a result of gravitational collapse. In that case the randomness conditions become much more complicated. Once again, it is not possible to invoke randomness at all the singularities. Black holes will swallow up the proverbial television sets, so viewed in reversed time their singularities throw out highly ordered states. The states in the vicinities of these multiple singularities will be causally connected in a complex network, and although the network as a whole may be chosen at random, the individual states cannot be.

3. CLASSICAL GRAVITATIONAL COLLAPSE AND BLACK HOLES

Although the most conspicuous entropy production in the Universe concerns the emission of starlight, which involves weak, strong and electromagnetic forces internal to the cosmological matter, there are also internal (non-global) gravitational forces which contribute to entropy production. These local gravitational effects take the form of density perturbations of the cosmological fluid, which also become important only after the fireball phase, when the temperature is sufficiently low to allow the steady growth of matter concentrations. The end products of these density perturbations are visible today as galaxies, stars and planets.

In classical statistical mechanics it is common to study a finite system enclosed inside a completely sealed box. A theorem due to Poincaré proves that the system will visit any given state consistent with the constraints, infinitely often, to within arbitrary accuracy. Classical statistical mechanics cannot be immediately extended to self-gravitating systems because of formidable difficulties of principle and technical complexity. Nevertheless, a number of basic principles can be deduced from heuristic arguments and *ad hoc* models.

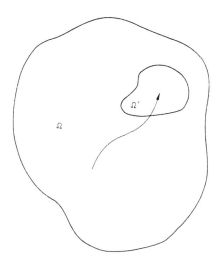

FIG. 1. *A schematic phase space Ω for a classical self-gravitating system. The region Ω' corresponds to catastrophic collapse to a black hole. Once the representative point crosses into Ω' it cannot leave. Quantum considerations suggest that the point can leave Ω' eventually, usually by passing out onto another sheet of Ω corresponding to a different total baryon number.*

If self-gravitation of the gas in the box is considered, the phase space of the system could contain a black hole. In I a schematic phase space Ω was introduced to describe the behaviour of a self-gravitating system (see Fig. 1). It is expected that the representative point for the box of gas will eventually visit the subset $\Omega' \subset \Omega$ which corresponds to the system being enclosed by an event horizon during catastrophic collapse. On its long route to Ω' this point would perhaps spend long periods of time in quasi-stable regions of Ω corresponding to stars, neutron stars etc. which represent local entropy maxima. They are not true equilibrium states, because of their inherent instability against black hole collapse.

The absence of a stable configuration for an object which has retreated inside its Schwarzschild radius (in the spherically symmetric case) suggests that the entropy of such an object is unbounded. Several heuristic arguments have been advanced in support of this conjecture. For example, the entropy change dS when a quantity of energy dE is added to a Schwarzschild black hole is formally given dE/T, where T is the black hole temperature. Because the hole is 'black', $T = 0$, so that the added entropy is formally infinite.

It can also be argued that the unchanging blackness of the hole when more energy is added implies that this energy behaves as though it were divided up into limitless new degrees of freedom, rather than enhancing the temperature in the existing degrees of freedom. These degrees of freedom may be used to store unlimited information, which is quite unobtainable from outside the black hole, on account of the no-hair theorems. The loss of unlimited information implies a divergent entropy.

In I the question was raised of statistical fluctuations in the internal states of a black hole. Because the black hole behaves like an object with an unlimited number of degrees of freedom, it would be expected that the macroscopic changes due to such fluctuations would be vanishingly small. Whereas for a non-gravitating system, ergodicity is expected, with the system revisiting all possible configurations, in Ω space this cannot occur. Every trajectory must cross the boundary of Ω' once and only once, because the formation of an event horizon prevents the system re-emerging from within the Schwarzschild radius. Thus Poincaré's theorem fails for a general classical self-generating system, and the collapse process is irreversible.

Towards the end of the last century, physics was faced with a closely analogous form of irreversible collapse in a purely attractive field of force. Two point particles with equal and opposite electric charge would, according to the time-reversible laws of classical electrodynamics, collapse onto each other without reaching a stable configuration, and in doing so create an unbounded quantity of entropy in the form of ever more energetic electromagnetic radiation. The paradox was removed, and a time-symmetric behaviour for charged particles recovered, by invoking the quantum theory.

It has long been conjectured (see Harrison *et al.* 1965) that quantum fluctuations by black holes would occur, causing both implosions and explosions. A quantum treatment of black holes would therefore be expected to damp out the entropy divergence (Bekenstein 1973).

4. EVAPORATING BLACK HOLES

The first person to give a detailed description of a quantum black hole was Hawking (1975). His great discovery revealed that an object undergoing gravitational collapse does not form a truly black hole at all, but radiates with a thermal spectrum at a finite temperature $T = \hbar K/2\pi k$ where K is the surface gravity, which is many orders of magnitude in excess of previous estimates (Harrison *et al.* 1965). The entropy of such an object turns out to be $S = Ak/4\hbar$ where A is the area of the event horizon ($= (4M)^{-1}$ for a Schwarzschild black hole of mass M). Thus as $\hbar \to 0$, $T \to 0$ and S diverges (units $G = c = 1$ have been used here).

In order to discuss the implications of the Hawking evaporation process for the question of time symmetry and fluctuations of black hole states, it is necessary to determine the effect of the back-reaction on the collapsing system due to the flow of radiation energy away from the black hole. Two dimensional calculations (see Davies 1976) indicate that a flux of negative energy occurs inside the collapsing matter, which flows inwards, across the horizon, at a rate precisely equal to the rate of energy loss from the Hawking radiation. The accumulation of negative energy inside the black holes has the effect of progressively reducing its total

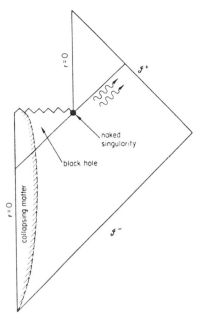

FIG. 2. *A classical Penrose diagram conjectured to describe an evaporating Schwarzschild black hole. After the evaporation a momentary naked singularity gives way to Minkowski space.*

mass, horizon area and entropy. As these quantities approach the Planck parameter, quantum gravity effects invalidate all existing treatments. Nevertheless Hawking (1975) has suggested that the end point of evaporation is described by the Penrose diagram shown in Fig. 2. Here the infalling matter strikes a singularity as in the non-quantum case but the radical departure in this case is the conjecture that when the horizon area shrinks to zero, the singularity is left naked, presumably to disappear (the mass has all been radiated away) leaving Minkowski space. The baryons which went to make up the collapse have disappeared also, an unpalatable circumstance blamed on the singularity, which has no obligation to obey the law of baryon number or, indeed, any other law.

Whether or not this scenario turns out to be correct, it is interesting to examine its implications for time asymmetry. The effect of the black hole evaporation process is firstly to enlarge the phase-space available to a system enclosed in a sealed box, by permitting changes in the total baryon number. Secondly, it clearly enables the conjecture of quasi-ergodicity to be reintroduced, because an initially dispersed system may collapse, and then evaporate to a dispersed state once again. However, unlike the conventional case of a non-gravitating ergodic system, the trajectories in Ω space are discontinuous, with end-points lying in a subset of Ω' corresponding to singular states.

To understand this remark, recall that ergodicity in a conventional system requires that the energy-surface in phase space be compact. The presence of singularities in a self-gravitating system violates the assumption of compactness. Information may pass to and from the system through the singularities. Inspection

185

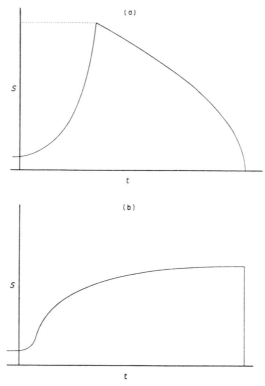

FIG. 3. *The entropy of a black hole as determined by (a) a distant observer, (b) an infalling observer. The figures are not to scale.*

of Fig. 2 shows that the collapsing matter enters a 'blind alley' region of the manifold and eventually intersects a singularity. The trajectory of this matter in Ω space cannot re-emerge from Ω'. We could regard Ω as the subspace of a much larger phase space which takes into account other manifolds coupled to the interior of the box 'through' the singularities, but for the purpose of observations inside the box the effect of matter passing into the blind alley is that the trajectory of this matter has an end point in Ω; the total system representative point then jumps discontinuously to a region outside Ω'.

The entropies of the collapsing matter as measured by (a) an observer far from the region of collapse (b) an observer falling in with the collapse, are compared in Fig. 3.

Observer (a) sees little of the entropy of the collapsing matter but can measure the entropy of the evaporation radiation. He measures a sudden (exponential) increase in this entropy as the collapse proceeds, leading to a momentary maximum of $S = 4\pi M^2$. Evaporation then proceeds to reduce the mass at an ever-increasing rate $[M \propto (t_0 - t)^{1/3},\ S \propto (t_0 - t)^{2/3}]$ so that the entropy of the object falls catastrophically to zero.

An observer who falls down the black hole sees little of the entropy of the annihilation radiation but can measure the entropy of the collapsing matter locally in a comoving frame. Conditions resemble a portion of a Friedmann universe as it recontracts to a final singularity. There will be some entropy growth due to the lagging of various processes behind the changing geometry, as described in Section 1, but these will progressively diminish as the temperature and density of the collapsing matter rise without limit on the approach to the singularity. This isentropic contraction is abruptly terminated when the singularity is reached, and 3N degrees of freedom suddenly disappear from the Universe. Thus, both observers agree on the initial and final entropy of the collapsing matter, but the intermediate details are quite different. To what extent the measurements of one may be related to that of the other is obscure, and will only be explained when a detailed theory of the the the statistical mechanics of strongly self-gravitating systems is well understood.

It was carefully explained in I that time asymmetry and irreversibility are quite distinct concepts. Consequently, the restoration of reversibility for self-gravitating systems as a result of Hawking's evaporation process does not imply that the behaviour of a black hole is time symmetric. The reverse process of evaporation requires the spontaneous appearance of N baryons at a naked singularity, to explode outwards in a white hole by drawing on energy provided by coherent beams of converging radiation, which annihilate in the gravitational field of the baryons. Because Ω space is non-compact, such an eventuality requires the cooperation of the singularities. Nothing is known about the physics in the vicinity of a singularity, so of course one is free to impose any discipline one likes on such an entity. Which rules are entirely a matter of taste. Hawking has suggested that singularities are at best subject to the fundamental principle of time reversal symmetry. If this is to be the case then the following argument ensues. Because black holes will form in the box only due to *random* rare fluctuations, the reverse white hole process must also be *random* if the time symmetry is to be preserved. That is, the matter which emerges from a white hole singularity must not contain correlations, which provides a very natural generalization of the cosmological randomness assumption. It is curious that black holes also possess the property that their emission quanta are uncorrelated. That is, it has been proved (Hawking 1975; Parker 1975; Wald 1976) that the expectation values of products of annihilation and creation operators which connect components of the density matrix of the final state of the quantum field near a black hole, containing different numbers of particles in the same mode or particles in different modes, are all zero. The absence of correlations in the black hole evaporation radiation has led Hawking (1976) to remark that black and white holes are indistinguishable. An attempt has been made to illustrate this in Fig. 4. Diagram (a) is the Penrose diagram for a black hole formed out of a converging flux of radiation. In a closed box, it is overwhelmingly probable that most of the energy is found in the form of black body radiation after an arbitrarily long time (the sides of the box are omitted on the diagrams for simplicity). A rare fluctuation causes the black hole to form. The length of the arrows indicates the intensity of the energy flux, which starts out large, but slowly falls and deviates from a Planck spectrum as the radiation in the box becomes depleted. The ingoing null lines are drawn extending on across the horizon as far as the singularity to indicate that this radiation continues inwards and hits the singularity.

The formation of the black hole causes Hawking radiation to be produced and flow out towards $\mathscr{I}^+$ along the outgoing null lines. These are drawn with

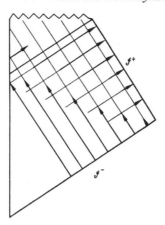

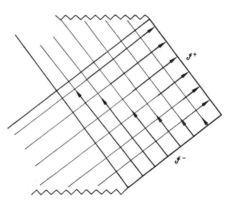

FIG. 4. *Penrose diagrams for (a) a random fluctuation resulting in Schwarzschild gravitational collapse, (b) the corresponding portion of the extended Kruskal manifold. When back-reaction is neglected, a distant observer cannot distinguish (a) from (b). As (b) is time symmetric this suggests that black and white holes are indistinguishable under these circumstances.*

end points among the collapsing (ingoing) radiation to indicate that the Hawking radiation is produced there. To a distant observer this radiation begins at low intensity and a non-thermal spectrum builds up to the Hawking value and Planck spectrum as the black hole forms. To this observer, the situation is indistinguishable from Fig. 4(b), where the radiation arises out of a white hole singularity in the past, crosses the past horizon and travels on out to $\mathscr{I}^+$. The essential point is that Fig. 4(b) is time-symmetric. If the arrows on the null rays are reversed, it may be inverted without change (other than an incidental change of $\mathscr{I}$ from right to left).

These considerations apparently become invalid when the effect of the back-reaction is taken into account. The back-reaction introduces correlations between quanta in different outgoing modes, which causes the spectrum to deviate from the thermal case. This is quite negligible until the very late stages of evaporation, when the black hole disappears in a burst of intensely hot radiation, well above the ambient temperature of the surroundings. The time reversal of this situation is for a black hole to form inside the box by a random fluctuation involving a small number of very energetic quanta from among the box radiation, whereas it is far more probable that such a fluctuation would arise from radiation at the average temperature inside the box. Put simply, most black holes end with a bang, but begin with a whimper. Exact time reversal symmetry is therefore absent. It follows that a white hole is not the same in detail as a black hole.

It will now be argued that the case of the closed box undergoing fluctuations to singular states does not suffer from the consistency problems encountered in the cosmological case in Section 1, when the randomness condition is applied to causally connected singularities. In the former case the singularities are imposed upon the system by the global dynamics of the Universe. This dynamics is not correlated with microscopic motions of the constituent particles, so that the final singularity in a recontracting Universe does not occur at random, but at a time which is causally determined by the initial big bang. In contrast, the system enclosed in a box is confined, not by its self-gravitation, *but by the walls of the box*. Consequently, the black and white holes form, and singularities occur, not as a result of global dynamics in accordance with the field equations of general relativity, but instead by the chance motions of the constituent particles which occasionally happen to assemble a portion of the contents of the box inside its Schwarzschild radius. In short, singularities *arise* at random in the box, so there is no inconsistency to conjecture that they *disperse* at random also.

ACKNOWLEDGMENTS

I should like to thank Dr S. A. Fulling and Professor R. Penrose for helpful discussions.

REFERENCES

Bekenstein, J. D., 1973. *Phys. Rev. D*, **7**, 2333.

Davies, P. C. W., 1974. *The physics of time asymmetry*, University of California press/Surrey University Press.

Davies, P. C. W., 1976. On the origin of black hole evaporation radiation. *Proc. R. Soc.*, in press.

De Witt, B. S., 1975. *Phys. Rep.*, **19**, 295.

Gold, T., 1962. *Am. J. Phys.*, **30**, 403.

Harrison, B. K., Thorne, K. S., Wakano, M. & Wheeler, J. A., 1965. *Gravitation theory and gravitational collapse*, chapter 11, University of Chicago Press.

Hawking, S. W., 1976. Black holes and thermodynamics, *Phys. Rev.*, **13**, 191.

Hawking, S. W., 1975. *Fundamental breakdown of physics in gravitational collapse*, California Institute of Technology, preprint.

Hawking, S. W., 1975. *Comm. Math. Phys.*, **43**, 199.

Landsberg, P. T. & Park, D., 1975. *Proc. R. Soc. A*, **346**, 485.

Layzer, D., 1976. *Astrophys. J.*, **206**, 559.

Parker, L., 1975. *Phys. Rev. D*, **12**, 1519.

Penrose, O. & Percival, I. C., 1962. *Proc. Phys. Soc.*, **79**, 605.

Penrose, R., 1974. Singularities in cosmology, *Confrontation of cosmological theories with observational data*, p. 263, ed. M. S. Longair, D. Reidel, Dordrecht, Holland.

Tolman, R. C., 1934. *Relativity, thermodynamics and cosmology*, Clarendon Press, Oxford.

Wald, R. M., 1975. *Comm. Math. Phys.*, **45**, 9.

Part D

Time in the Arts

193

IX. *A broader view of life*. A Dance to the Music of Time, by Nicolas Poussin (1638–40). One view of the picture is that the two females on the right are Wealth and Poverty, Wealth willing just to touch the hand of Poverty (Poverty wearing a linen head-dress and Wealth wearing a string of pearls). The two other females are presumed to be Pleasure and Fame. Other views are that the figures represent Poverty, Labour, Riches and Luxury. The double face of Terminus looks into both past and future. Phoebus in his golden car is attended by the Hours and is preceded by Aurora. Time plays the lyre while children play with the hour glass. The wheel of Fortune goes round and round as in a dance of human life to which all can contribute.

'These classical projections . . . suddenly suggested Poussin's scene in which the Seasons, hand in hand and facing outward, tread in rhythm to the notes of the lyre that the winged and naked greybeard plays. The image of Time brought thoughts of mortality: of human beings, facing outward like the Seasons, moving hand in hand in intricate measure: stepping slowly, methodically, sometimes a trifle awkwardly, in evolutions that take recognisable shape: or breaking into seemingly meaningless gyrations . . . unable to control the melody, unable, perhaps, to control the steps of the dance'.

(From Anthony Powell, *A Question of Upbringing* (1951, London: Heinemann).)

194

MOMENT AND MOVEMENT IN ART

By E. H. Gombrich

While the problem of space and its representation in art has occupied the attention of art historians to an almost exaggerated degree, the corresponding problem of time and the representation of movement has been strangely neglected. There are of course some relevant observations scattered throughout the literature,[1] but no systematic treatment has ever been attempted. It is not the purpose of the present paper to supply this want: only to indicate how this neglect may have arisen and where we may have to revise our preconceptions if we are to approach the problem afresh. For it may be argued that it was the way in which the problem of the passage of time in painting was traditionally posed that doomed the answers to relative sterility. This tradition reaches back at least to the early eighteenth century, more precisely to Lord Shaftesbury's classic formulation in the *Characteristics*.[2] Chapter I of *A notion of the Historical Draught, or Tablature of the Judgment of Hercules* opens with the statement that 'this Fable or History may be variously represented, according to the Order of Time'

> Either in the instant when the two Goddesses (VIRTUE and PLEASURE) accost HERCULES; Or when they are enter'd on their Dispute; Or when their Dispute is already far advanc'd, and VIRTUE seems to gain her Cause.

In the first instance Hercules would have to be shown surprised at the appearance of the two Goddesses; in the second he would have to be shown interested and in doubt, and in the third we would witness how he 'agonizes, and with all his Strength of Reason endeavours to overcome himself'. It is this Aristotelian turning-point that is recommended to the painter, though Shaftesbury also discusses the fourth possibility of representing 'the Date or Period . . . when Hercules is intirely won by Virtue'. He rejects it on the grounds of dramatic inefficacy and for the additional reason that in such a picture 'PLEASURE . . . must necessarily appear displeas'd, or out of humour: a Circumstance which wou'd no way sute her Character.'

> 'Tis evident, that every Master in Painting, when he has made choice of the determinate Date or Point of Time, according to which he wou'd represent his History, is afterwards debar'd the taking advantage from any other Action than what is immediately present, and belonging to that single Instant he describes. For if he passes the present only for a moment, he may as well pass it for many years. And by this reckoning he may with as good right repeat the same Figure several times over . . .

[1] Cf. E. Panofsky's review of Hanns Kauffmann, *Albrecht Dürer's rythmische Kunst* in *Jahrbuch für Kunstwissenschaft*, 1926. M. J. Friedlaender, *Von Kunst und Kennerschaft*, Oxford and Zurich, 1946, pp. 60–66; H. A. Groenewegen-Frankfort, *Arrest and Movement*, 1951; H. van de Waal, *Traditie en beziehung*, Rotterdam, 1946 and 'De Staalmeesters en hun legende', *Oud Holland*, 71, 1956; R. Arnheim, *Art and Visual Perception*, 1956, chapter VIII, and my *Art and Illusion*, 1960, index s.v. 'movement'.

[2] Anthony, Earl of Shaftesbury, *Characteristicks of Men, Manners, Opinions, Times*, 1714. For the commission and its result see F. Haskell, *Patrons and Painters*, 1963, p. 198.

There remains no other way by which we can possibly give a hint of any thing future, or call to mind any thing past, than by setting in view such Passages or Events as have actually subsisted, or according to Nature might well subsist, or happen together in *one and the same* instant.

This absolute necessity, however, need not prevent the painter from representing movement or change such as the turning-point of the drama Shaftesbury had recommended for the choice of Hercules. For 'the Artist has power to leave still in his Subject the Tracts or Footsteps of its Predecessor . . . as for instance, when the plain Tracks of Tears new fallen . . . remain still in a Person newly transported with Joy. . . . By the same means, which are employ'd to call to mind *the Past*, we may anticipate *the Future* . . .' In our case, for instance, the artist could show Hercules in doubt and yet indicate that his decision was to be in favour of Virtue:

> This Transition, which seems at first so mysterious a Performance, will be easily comprehended, if one considers, That the Body, which moves much slower than the Mind, is easily out-strip'd by this latter; and that the Mind on a sudden turning itself some new way, the nearer situated and more sprightly parts of the Body (such as the Eyes and Muscles about the Mouth and Forehead) taking the alarm, and moving in an instant, may leave the heavier and more distant parts to adjust themselves, and change their Attitude some moments after. This different Operation may be distinguish'd by the names of *Anticipation* and *Repeal*.

Shaftesbury admits that this rigorous standard of instantaneous action is often sinned against. He refers with amusement and contempt to the usual representations of Diana and Actaeon, in which the goddess is seen throwing water at Actaeon whose horns are already growing although he is not yet wet.

It was Shaftesbury's formulation, no doubt, which influenced James Harris in his *Discourse on Music, Painting and Poetry*,[3] where the distinction is first made with all desirable clarity between the various media of art, music being concerned with motion and sound, painting with shapes and colours. Every picture is thus 'of necessity a *punctum temporis* or instant'. But though Harris calls a painting 'but a Point or Instant', he adds that 'in a Story well known the Spectator's Memory will supply the previous and the subsequent . . . [This] cannot be done where such Knowledge is wanting'. In fact he wonders whether any historical incident in a painting would be intelligible 'supposing history to have been silent and to have given no additional information'.

All these ideas were taken up by Lessing and woven into the fabric of his *Laocoon*, which systematically distinguishes between the arts of time and the arts of space. 'Painting can . . . only represent a single moment of an action and must therefore select the most pregnant moment which best allows us to infer what has gone before and what follows.'[4] Lessing, as I have tried to argue elsewhere,[5] did not write the Laocoon for the sake of this well-estab-

[3] James Harris, *Three Treatises*, 1744.
[4] *Loc. cit.*, xvi.
[5] Cf. my 'Lessing' (Lecture on a Master Mind), *Proceedings of the British Academy*, xliii, 1957.

lished distinction. What provoked him was the idea that poetry or drama should ever conform to the limitations of the visual arts. For these limitations, he thought, followed precisely from the restriction to one single moment. If there is to be one moment that will be transfixed and preserved for eternity it clearly must not be an ugly moment. The famous disquisition about the reasons why the marble Laocoon must not shout while Vergil can let his Laocoon groan and bellow is deduced from this *a priori* principle.

> The artist can never use more of ever-changing reality than one single moment of time and, if he is a painter, he can look at this moment only from one single aspect. But since their works exist not only to be seen but also to be contemplated, contemplated at length and repeatedly, it is clear that this single moment and single aspect must be the most fruitful of all that can be chosen. Only that one is fruitful however that gives free rein to the imagination. The more we see, the more we must believe ourselves to be seeing. There is no moment however in the whole sequence of an emotion which enjoys this advantage less than its climax. Beyond it there is nothing and thus to show the eye the extreme, means to clip the wings of the imagination. . . . Thus when Laocoon sighs the imagination can hear him shout, but when he shouts our mind can neither rise to greater intensity nor descend to a lower step without picturing him in a more tolerable and therefore less interesting state.[6]

Unconvincing as this casuistry may be, it was meant as a concession to Winckelmann who had never failed to denounce the arch-corrupter Bernini, in whose works such as the *David* or the *Anima Dannata* the climax of movement and passion is indeed presented to the eye. As long as the arts of time remained free to depict these extremes, Lessing was quite ready to concede that the visual arts should concentrate on the moments of stillness instead.

These particular conclusions were implicitly challenged by the Romantics,[7] but as far as I know the underlying distinction between the art of time and of space, of succession and simultaneity, remained unquestioned in aesthetics. Thus the artist was driven in the interest of truth to concentrate more and more on the task of giving, in Constable's words, 'One brief moment caught from fleeting time a lasting and sober existence'. These words were written in 1832.[8] A few years later photography was invented. But the early photograph with its long exposure time was not yet a threat to the artist who set himself the aim of catching time on the wing. When Ruskin wrote his chapter in *Modern Painters*, 'Of Truth of Water' to exalt the fidelity of Turner's renderings over the earlier conventions of Van der Velde or Canaletto, he regrets that he 'cannot catch a wave, nor Daguerreotype it, and so there is no coming to pure demonstration'.[9] However, he was clearly convinced that the Daguerreotype would prove Turner right if it ever could catch a wave.

And yet when the camera did finally catch up, it appeared to demonstrate

---

[6] *Loc. cit.*, III.
[7] Cf., e.g., Friedrich Schlegel's defence of the subject of martyrdoms in paintings, *Gemaeldebeschreibungen aus Paris und den Niederlanden, II. Nachtrag* (1804).

[8] J. Constable, *Various Subjects of Landscape*, 1832. On Constable and photography see the note by R. Beckett, *infra*, p. 342.
[9] *Loc. cit.*, section V, chapter I.

the inferiority even of the most sensitive eye. The notorious issue over which the battle broke was the rendering of the galloping horse.[10] The photographer Muybridge, in 1877, went to great trouble to solve the problem of what really goes on in this rapid movement. He lined up twelve cameras along a race-course in California in such a way that the passing horse would break a thread stretched across its path and thus release the shutter. The dazzling sun of California allowed a brief exposure, and in 1878 Muybridge could startle the world of art and of science with his demonstration that painters could not see. In particular the flying gallop, so frequent in the rendering of horse races, was claimed to be quite at variance with the facts. The reaction of painters and critics was ambivalent. Some said that it was the instantaneous photo-graph that looked unreal and that the experiment had proved the superiority of art. They pointed to the strangely frozen effect of instantaneous photo-graphs. It is hard for us to recapture the puzzled curiosity which these once caused. We see so many pictures of football matches and athletic events in our papers that we have come to take these chaotic configurations for granted. Only once in a while does an action photograph really puzzle us with an impression of impossible movements. On the whole it is far from true that all snapshots look frozen to us. No wonder artists wanted to accept the challenge of the camera and tried to learn from it, thus endorsing the traditional view that the truthful image can or should only render what we actually manage to see in a moment.[11]

Now there certainly is a sense in which the instantaneous photograph represents the truth of that moment. Put a succession of snapshots taken at quick intervals into a revolving drum so that each is visible through a slot for about one-sixteenth of a second and we see the original event in motion. Thanks to this convenient illustration we can in fact pose Shaftesbury's pro-blem in a very simple way. Suppose a news camera had filmed the Judgement of Hercules. Which of the frames would be suitable for publication as a still from the film? The answer is that none might do. The so-called 'stills' which we see displayed outside cinemas and in books on the art of the film are not, as a rule, simply isolated frames from the moving picture enlarged and mounted. They are specially made and very often specially posed on the set, after a scene is taken. That thrilling scene where the hero embraces his girl while he keeps the villain covered with a revolver may consist of many yards of film containing twenty-four frames per second of running time, but not one of them may be really suitable for enlargement and display. Legs fly up in the air, fingers are spread out in an ungainly way and an unintelligible leer comes over the hero's face. Far better to pose the scene carefully and photo-graph it as a readable entity which fulfils Shaftesbury's and Lessing's demands for anticipation and repeal, though the posed stills partly refute the theory that a real *punctum temporis* will easily combine all the necessary cues in one simultaneous assembly.

I do not want to overstate the force of this particular refutation. There

[10] Beaumont Newhall, 'Photography and the development of kinetic visualization', this *Journal*, VII, 1944, pp. 40–45. See also S. Reinach, *Le représentation du Galop dans l'art*

ancien et moderne, 1925.

[11] I am indebted to Mr. Aaron Scharf who is preparing a book on this problem.

are stills which are taken from the film frames and are perfectly legible, just as there are instantaneous photographs which do give us the perfect illusion of a coherent action. Conceivably the struggle of a man and his two sons with monstrous serpents might pass through the configuration of the Laocoon. But could anyone ever know this for certain? Do we not beg the most important question when we ask what 'really happens' at any point of time? We therewith assume that what Harris called a *punctum temporis* really exists, or, more radically, that what we really perceive is the infinite sequence of such static points in time. Once this is conceded the rest follows, at least with the demand for mimesis. Static signs, the argument runs, can only represent static moments, never movements which happen in time. Philosophers are familiar with this problem under the name of Zeno's paradox, the demonstration that Achilles could never catch up with a tortoise and no arrow could ever move.[12] As soon as we assume that there is a fraction of time in which there is no movement, movement as such becomes inexplicable.

Logically the idea that there is a 'moment' which has no movement and can be seized and fixed in this static form by the artist, or for that matter, by the camera, certainly leads to Zeno's paradox. Even an instantaneous photograph records the traces of movement, a sequence of events, however brief. But the idea of the *punctum temporis* is not only an absurdity logically, it is a worse absurdity psychologically. For we are not cameras but rather slow registering instruments which cannot take in much at a time. Twenty-four successive stills in a second are sufficient to give us the illusion of movement in the cinema. We can see them only in motion, not as stills. Somewhere along this order of magnitude, a fifteenth or a tenth of a second, lies what we experience as a moment, something we can just seize in its flight. Compared with the speed of a computer we are indeed slow in the uptake.

The television screen is an even more impressive demonstration of this slowness of our perception and the duration of what we consider a 'moment'. When we watch the programme we are, in fact, watching a tiny spot of light traversing the screen from side to side 405 times in one-fifth of a second at a speed of about 7,000 miles an hour. This spot of light traces out the rectangular area on which the picture is seen. The camera scans the object with this beam which varies in intensity as it strikes brighter or darker objects, and these fluctuations are translated into electrical impulses and re-translated into a travelling scanning beam in the television set. At each moment of time, therefore, what we really see (if that expression had any meaning) would only be one luminous dot.[13] It could not even be called a brighter or less bright dot, since these notions introduce what has gone before and what comes after. It would be a meaningless dot. Actually if we want to pursue this thought to its logical conclusion the *punctum temporis* could not even show us a meaningless dot, for light has a frequency. It is an event in time, as is sound—not to speak of the events in the nervous system that transform its impact into a sensation.

These considerations may allow us to focus more sharply the philosophical problem that underlies the traditional distinction between the arts of time and the arts of space. As a process in time television certainly presents the

[12] My attention was drawn to this connection by Dr. William Bartley III.

[13] Donald G. Fink and David M. Lutyens, *The Physics of Television*, 1961.

20

travels of a meaningless flickering dot, and the extension in space of this dot turns out to be an illusion, founded on the sluggishness of our perception. Yet it is this sluggishness that overcomes the limitation of time, the *punctum temporis*, and creates a meaningful pattern through the miracle of persistence and memory.

It was no less a thinker than St. Augustine who pondered this miracle in one of the most famous meditations of his *Confessions*.[14] Famous, but still not sufficiently so. For if Shaftesbury and Lessing had profited from the lesson of St. Augustine's introspections they could not have created that fatal dichotomy between space and time in art which has tangled the discussion ever since.

What puzzles St. Augustine is precisely the elusiveness of the present moment flanked as it is by future time that is not yet, and past time that is no longer. How can we speak of the length of time, how can we even measure time since what we measure is either not yet or no longer in existence? It is Zeno's paradox from a new angle. For we do speak of long times, says St. Augustine, we also speak in poetics of long and short syllables, and everyone knows what we mean when we say that a long syllable in a poem is double the length of a short one. And yet I can only call the syllable long after it has ended, when it no longer *is*:

> What is it therefore that I measure? Where is that short syllable by which I measure? Where is that long one which I measure? Both have sounded, have flown and gone, they are now no more: and yet I measure them . . . it is not these sounds, which are no longer, which I measure, but something that is in my memory that remains fastened there. It is in thee my mind that I measure my times. Please do not interrupt me now, that is do not interupt thine own self with the tumults of thine own impressions. In thee, I say, it is, that I measure the times. The impression, which transient things cause in thee and which remains even when they have gone, that is it which being still present I measure.[15]

And as with the past, so with the future. 'Who can deny that things to come are not yet? Yet already there is in the mind an expectation of things to come.'[16]

And then comes the famous introspective account of what happens in his mind when he recites a psalm.

> Before I begin my expectation alone extends itself over the whole, but so soon as I shall have once begun, how much so ever of it I shall take off into the past over so much my memory also reaches, thus the life of this action of mine is extended both ways: into my memory, so far as concerns the part I have repeated already, and into my expectation too, in respect of what I am about to repeat.[17]

When Professor Hearnshaw gave the presidential address at the British Psychological Society in 1956[18] he made this passage the starting-point for

---

[14] Book XI, 10–31.

[15] *Ibid.*, XI, 27; the translation follows the one by William Watts (1631) used in Loeb Classical Library, 1912.

[16] XI, 28.

[17] XI, 28.

[18] *Bulletin of the British Psychological Society*, 30 September 1956.

the discussion of what is technically known as 'temporal integration', the bundling together in one extended stretch of time of memories and expectations. Even in his own field he noticed a scarcity of literature on this all-pervasive problem, particularly in comparison with the 'extraordinary dominance of special concepts, notably in Gestalt psychology'. His explanation applies with equal force to our own field of study: 'Temporal integration cuts across faculty boundaries. It implies perception of the present, memory of the past, and expectation of the future—stimulus patterns, traces and symbolic processes—integrated into a common organization.' It is the kind of complex problem that research shies away from.

Not that the last hundred years have not yielded insights which allow us to pose St. Augustine's problem with more precision, though scarcely with the same beauty. We know for instance that where he speaks of memory that retains the present in the mind, we can and must distinguish between at least three types of such retention. The first is that persistence of a sense impression that is so relevant to television. This is, partly at least, a physiological process which makes the impression of light and sound persist for a moment when the actual stimulus is over. But apart from this, there is another kind of persistence or reverberation that is variously known as 'immediate memory' or 'primary retention'.[19] It is in an elusive concept, but one easily open to introspection. It happens that somebody says a few words which we fail to take in. But as we cast our mind back we find that the sound is still there, a few seconds later, and we can find out what the words meant. This kind of immediate memory is a trace that disappears very quickly, but it is vital for our real understanding of St. Augustine's problem of what we measure when we measure the lengths of syllables in a poem or of tones in a melody just heard. These syllables or tones are still really there in a rather different sense from which things past are still stored somewhere in our mind. Indeed Hebb, in his book on the *Organization of Behavior*,[20] postulates two distinct kinds of memory. He would like to believe in some kind of reverberation of the stimulus which carries the memory until a more permanent trace is formed. Be that as it may, it really is evident that our impressions remain available for a brief span of time, the time that is known as the memory span or the specious present. Psychological experiments with the memorizing of nonsense syllables or digits show that subjects can hold a limited number over a few seconds after which they vanish and are replaced by fresh incoming impressions. There is a fascinating paper by G. A. Miller called 'The Magical Number Seven plus and minus 2'[21] in which he puts forward the hypothesis that seven acquired its status precisely as the largest number of items we can generally hold at once. What matters to us in these experiments and speculations is that they corrode the sharp *a priori* distinction between the perception of time and of space. Successive impressions do in fact persist *together* and are

[19] *The Third Annual Report of the Center for Cognitive Studies at Harvard*, 1963, p. 142, contains a preliminary account of a study of this phenomenon under the graphic name of the 'echo box'.

[20] D. O. Hebb, *The Organization of Behavior*, 1949, pp. 61, 62.
[21] Cf. D. C. Beardslee and W. Wertheimer, *Readings in Perception*, 1958.

not wholly experienced as successive. Without this holding operation we could not grasp a melody or understand the spoken word.

St. Augustine was right, moreover, when he found that the mind not only retains the past impression but also reaches out into the future. And again, more is involved here than a mere expectation extending over any length of time. Take St. Augustine's example of the recitation of a psalm. While we speak one line we are in a real sense making ready for the recital of future lines. Lashley has shown in a classic contribution to this subject[22] that the immediate future we are thus making ready for is as much really present in our mind as is the past. If it were not so spoonerisms would not occur. When Dr. Spooner said to a student 'You have hissed my mystery lesson' he proved that the letters to be spoken were already present together in his mind and so got mixed up. One of Lashley's most interesting examples concerns typing mistakes. It happens for instance that we double the wrong letter in a word we perfectly know how to spell. The instruction to double, which is already waiting in the wings, as it were, misses its cue and is applied wrongly.

Here, where our own actions are concerned, we can speak of innervations stored in advance ready to go into action in some predetermined serial order or sequence—and surely it is much less surprising that this goes wrong occasionally than that it ever comes right. But even if we do not speak but listen to speech, if we do not play but listen to music, some representation of possibilities to come will be stored in readiness just present to be triggered off by the slightest confirmatory cue. There is the story of the unfortunate singer who discovered that he could not sing the highest note in an aria. So he stepped forward and opened his mouth widely and triumphantly while the orchestra made a loud noise. The public 'heard' the top note and applauded. The note was as much present in the public's mind at that moment as were the notes that led up to it. What would have happened if the singer had sung a false note instead? He would not only have hurt our ears in that moment, he would retroactively have spoilt the whole phrase even if the earlier notes had been provisionally classified and stored as acceptable.

Experiences of this kind illustrate why the old distinction between the arts of time such as music and poetry, and the arts of space such as painting and architecture, is so barren and misleading. In listening to music the moment is as it were spread out to a perceptual span in which immediate memory and anticipation are both phenomenally present. Some people feel this presence strongly as a spatial pattern, others less so. It hardly matters as long as we recognize that the understanding of music or the understanding of speech would not be possible if we lived in too narrow a present. For in music no less than in speech what comes after affects what has come before. There is a blasphemous musicians' joke attributed to Arthur Schnabel showing how the most heavenly theme, such as the opening of Mozart's G minor symphony, can be fouled and destroyed by the simple doubling of the last note of the phrase. The same can be demonstrated in the hearing of speech where we really could not make sense of any sentence unless we scanned the recent sounds and revised our interpretation according to the way our expec-

[22] 'The Problem of Serial Order in Behavior', *Cerebral Mechanisms in Behavior*, ed. L. A. Jeffres, New York, 1951.

tations are confirmed or refuted by the next sound. Lashley gives two amusing examples for this influence of context on meaning: 'The mill-wright on my right thinks it right that some conventional rite should symbolize the right of every man to write as he pleases.' Here it is mainly the preceding words which influence our pigeonholing. But in his other instance the retroactive process comes into its own: 'Rapid righting with his uninjured hand saved from loss the contents of the capsized canoe.' The associations which give an unexpected meaning to the sound 'righting', as Lashley says, are not activated for at least three to five seconds after hearing the word. But as a rule the word is still present for this revision.

Psychological time is clearly something much more complicated and mysterious than the sheer succession of events. But if music or poetry are not so exclusively arts of succession as Shaftesbury, Harris and Lessing held, painting or sculpture are not as clearly arts of arrested movement. For, phenomenologically, that moment does not exist for the painter any more than it exists for the musician. If in hearing we assemble our impression in some kind of short-term storage, before we confine them to memory proper, we do and have to do the same thing in seeing. Visual perception itself is a process in time, and not a very fast process at that. Measurements have been made about the amount of information the eye can take in at a glance and attempts have been made, especially by the late Prof. Quastler, to give precision to these two concepts. His conclusion was that we generally vastly overrate the amount of information we process. 'What we actually see is a very rough picture with a few spots in clear detail. What we feel we see is a large picture which is everywhere as clear in detail as the one favourite spot on which we concentrate our attention. Roughly speaking the area of clear perception includes less than one per cent of the total visual field.'[23] We might add that the existence of the macula, the blind spot, was only discovered relatively late. Why? Because we can scan our surroundings for information and retain the result of previous scannings together with the anticipations of future impressions which can become critically important in confirming or revising a percept.

In that sense it is surely true to say that we never see what the instantaneous photograph reveals, for we gather up successions of movements, and never see static configurations as such. And as with reality, so with its representation. The reading of a picture again happens in time, in fact it needs a very long time. There are examples in psychological literature of the weird descriptions given by people of identical paintings flashed on to a screen for as long as two seconds.[24] It takes more time to sort a painting out. We do it, it seems, more or less as we read a page, by scanning it with our eyes. Photographs of eye movements suggest that the way the eye probes and gropes for meaning differs vastly from the idea of the critics who write on the artist 'leading the eye' here or there.[25] Not that these aesthetic experiences need

[23] Henry Quastler, 'Studies of Human Channel Capacity', *Control Systems Laboratory Report*, Number R.–71, p. 33 of the report (circulated in stencilled form).
[24] F. C. Bartlett, *Remembering*, 1932, pp. 29 ff.; M. D. Vernon, *A Further Study of Visual Perception*, 1952, Appendix B.
[25] J. J. Gibson, *The Perception of the Visual World*, 1950, p. 155.

be entirely spurious or fictitious. They may be *post factum* reconstructions facilitated by retroactive revision. The most extreme and therefore the most instructive case of such retroaction on past sensations concerns the 'eidetic faculty'.[26] Eidetic children are able to inspect a picture for a few seconds and then to visualize it as on a screen even though it has been withdrawn. They can for instance subsequently read an inscription over a door or count the chickens in the yard even though they had not done so during the exposure of the picture. But it has been shown that their memory is not simply a photographic reproduction. Many pictures portraying action result in an image where the action is carried to completion. 'In imaging a picture containing a donkey standing some distance from a manger, the donkey crossed over to the manger, moved its ears, bent its neck, and began to eat. Suggestions from the experimenter that the donkey was hungry sometimes served to set in motion a series of changes that surprised the imaging children themselves. It was as if they were not now looking at a static picture but at a living scene, for, as soon as the suggestion was given, the donkey would spontaneously race over to the manger.' (After H. Kluver, 1926.)

We have here an almost pathological magnification of what goes on in all of us when we look at a picture. We build it up in time and hold the bits and pieces we scan in readiness till they fall into place as an imaginable object or event, and it is this totality we perceive and check against the picture in front of us. Both in hearing a melody and in seeing a representation what Bartlett called the 'effort after meaning' leads to a scanning backward and forward in time and in space, the assignment of what might be called the appropriate serial orders which alone give coherence to the image.

In other words, the impression of movement, like the illusion of space, is the result of a complex process which is best described by the familiar term of reading an image. It cannot be the purpose of the present paper to explore this process afresh.[27] But one principle that applies to the reading of spatial relationships on a flat canvas can easily be shown to apply no less to the reconstruction of temporal relationships. It may be called the principle of the primacy of meaning. We cannot judge the distance of an object in space before we have identified it and estimated its size. We cannot estimate the passage of time in a picture without interpreting the event represented. It is for this reason perhaps that representational art always begins with the indication of meanings rather than with the rendering of nature and that it can never move far from that anchorage without abandoning both space and time. What else is the so-called 'conceptual image', the primitive pictograph of the child or the untutored, than the assertion of this primacy? A letter sent by the chieftain of an American Indian tribe to the President of the United States illustrates this principle (Pl. 29b).[28] He is seen extending the hand of peace to the man in the White House. The indications of other creatures and huts signify that members of his totem and other tribes are now ready to live in houses and give up the life of nomads. Strictly speaking, then, there is no moment of time represented here—as little indeed as there is a real space in

[26] For the following see Ian M. L. Hunter, *Memory, Facts and Fallacies*, 1957, pp. 148–9.
[27] Cf. my *Art and Illusion, passim.*

[28] W. Wundt, *Völkerpsychologie*, I. 1, 1911, p. 247.

a—Raising of Lazarus. Mosaic. Ravenna, S. Apollinare Nuovo

b—American Indian picture-letter. From W. Wundt, *Völkerpsychologie*, I, i, p. 247

d—Scene from *Los Olvidados*. New York, Museum of Modern Art, Film Library

c—Sebastiano del Piombo, *Raising of Lazarus*. London, National Gallery

e—Death of Orpheus. From Ovid, *Metamorphoses*, Lyons, 1507, fol. 152

a—Giotto, *Presentation of the Virgin.* Padua, Arena Chapel.
b—Ghirlandajo, *Presentation of the Virgin.* Florence, S. Maria Novella

c—Titian, *Presentation of the Virgin.* Venice, Accademia

d—Tintoretto, *Presentation of the Virgin.* Venice, Madonna dell'Orto

which the surrender takes place. And yet it is the gesture of the extended hand which ties the individual pictographs together into one coherent message and meaning. This meaning could be re-enacted in a real ceremony or represented in a realistic picture, but in every case it would have to centre on the chieftain's gesture or its symbolic equivalent which alone could convey to us that first there was war and now there is peace.

The narrative art of all periods has made use of such symbolic gestures to convey the meaning of an event. Take the illustration in S. Apollinare Nuovo in Ravenna of the raising of Lazarus reduced to its essential elements—the figure of Christ, the mummy in the tomb, and the gesture of power that effects the change (Pl. 29a). But is it different with a more ambitious rendering, such as Sebastiano's staging of the same event? (Pl. 29c). Is it not all assembled round the gesture that makes the meaning cohere?

In a sense this appears to confirm Shaftesbury's and Lessing's idea that the successful illustration of a narrative will always suggest and facilitate repeal and anticipation, the scanning backward and forward in time that comes from the understanding of an action. But we can also see more clearly why Shaftesbury was mistaken when he wrote that 'every painter when he has made his choice of the determinate Date or Point on Time . . . is afterwards debar'd taking advantage from any other action than what is immediately present and belonging to that single instant he describes. For if he passes the present only for a moment, he may as well pass it for many years.' There is a real difference between events assembled in one memory span and subordinate to one central perceptible meaning and events separated by years or even by hours. Just as music unfolds in phrases, so action unfolds in phases, and it is these units which are somehow the experienced moments in time, while the instant of which the theoreticians speak, the moment when time stands still, is an illicit extrapolation, despite the specious plausibility which the snapshot has given to this old idea.

If we ask ourselves what quality a snapshot must possess to convey the impression of life and movement we will find, not unexpectedly, that this will again depend on the ease with which we can take in the meaning that allows us to supplement the past and arrive at an anticipation of the future. It must be the same with stills from films. A scene such as the extract from *Los Olvidados*[29] (Pl. 29d) is only too clear because we understand the logic of the situation, the threatening posture of the boys and the protective gesture of the victim. It is well-known that this configuration has been often repeated in art in the context of battle-scenes and such events as the killing of Orpheus (Pl. 29e). Did the producer of the Mexican film derive this formula from the classical *Pathosformel* which so interested Warburg?[30] Hardly. There are few other ways in which this meaning could be conveyed with such ease.

It would be interesting sometimes to ask oneself what objective timespan is gathered together in this way by the meaning conveyed. Take the iconography of the Presentation of the Virgin: Giotto shows us St. Anne actually leading the Virgin up the step directly into the care of the High Priest (Pl. 30a). The import of the action is emphasized by that dramatic device

[29] S. Kracauer, *The Nature of Film*, 1961, fig. 25.

[30] A. Warburg, 'Dürer und die italienische Antike' (1905), *Gesammelte Schriften*, ii, 1932.

of bystanders not looking at the scene itself but at each other, which extends the time span. They have seen what is happening and are now exchanging glances or remarks. In Ghirlandajo (Pl. 30b) the distance the Virgin has to traverse from her family to the waiting high priest is larger, and so is the assistant crowd. The span increases in Titian's composition, but the gestures are the same (Pl. 30c). Tintoretto (Pl. 30d) changes the direction of the path and the intensity of the reaction among the beggars and cripples, but he too relies on the pointing action and the High Priest's gesture of welcome.

It is customary to describe this last type of composition as less static, more restless, *mouvementé* or *bewegt* than the earlier examples which strike us as comparatively calm, and possibly even posed as in a 'still'. The violent movement of some of the figures and particularly their instability clearly contributes to this impression that a moment has here been caught that could not have lasted more than a split second. But we also feel that the composition itself, the comparative complexity of the arrangement and the steep curve of the steps, enhances this reaction. After all we might easily describe the architecture itself in terms like 'dynamic', as if we experienced the forms to be in motion. Though these belong to the commonplaces of criticism, it is not quite easy fully to account for this terminology. Why is symmetry experienced as static, asymmetry as unstable; why is any lucid order felt to express repose, any confusion movement?

It is unlikely that there is one cause underlying these reactions or, indeed, that they are not at least partially conditioned by cultural conventions. But one would guess that here, as so often, the metaphors we use can guide us at least some way. Balanced objects can remain static where lopsided ones will fall any moment, and so the tendency is to seek for the reassuring balance and to expect a rapid change where it is absent. It is curious how easily this experience is transferred, as if by analogy, to other configurations. A leaflet for amateur photographers[31] (Pl. 31a) rightly points out that a sailing-boat photographed in the centre of an opening will look becalmed, one shown off-centre will appear to move. Of course this applies with much greater force to sailing-boats than, for instance, to trees, which suggests that even here meaning has a large share in the resultant impression.

Even so we seem to be presented with a strange paradox—the understanding of movement depends on the clarity of meaning but the impression of movement can be enhanced by lack of geometrical clarity. The most interesting test case here is an experiment made by Donatello. The dancing *putti* of his Prato pulpit (Pl. 31d) are gay and sprightly enough, but when the master came to develop the idea in the *Cantoria* (Pl. 31c) he deliberately obscured their dance by the daring device of placing it behind a row of columns. For most observers the effect of turbulent movement is enhanced by this partial masking. Could it be that another analogy contributes to this effect? That the difficulty we experience in following and integrating the scene fuses with memories of the difficulty we might experience in reality in sorting out the bodies and limbs of a whirling dance? In both cases, after all, the eye would send back to the brain the message 'hard to catch' and so the two might be interchangeable. Maybe Donatello consciously or unconsciously exploited an

[31] 'Auf das Sehen kommt es an', issued for Ilford by Ott & Co., Zofingen.

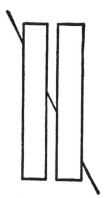

a—Setting and movement

b—Poggendorf illusion

c—Donatello, *Cantoria* (detail). Florence, Museo del Duomo

d — Donatello, Pulpit. Prato, Cathedral

e—*Discobolos*. Munich, Antiquarium. Front view

f—*Discobolos*. Munich, Antiquarium. Side view

209

a—Picasso, *Sleeper turning*. Zervos, *Picasso*, XI, 1960, fig. 198. Courtesy Editions Cahiers d'Art, Paris

b—Picasso, *Girl reading*. London, Private collection

c—Bridget Riley, *Fall*, 1963. London, Tate Gallery

additional effect that arises out of such masking—the effect known to psychologists as the Poggendorf illusion (Pl. 31b). A line that appears obliquely to pass behind a band or rectangle frequently looks as if its continuation were displaced. And so the possibility exists that the leg of the child really seems to have moved while we scan the composition. There must be yet other reasons while incompleteness can contribute to the impression of rapid movement. The 'snapshot effects' of Degas sometimes give the impression that the artist was so intent on fixing a motif on the canvas that he had neither time nor opportunity of seeking an advantageous viewpoint. The incompleteness becomes an indication of the painter's hurry, his own preoccupation with time which is contagious. We, too, speed up our scanning, and it can happen in the process that the incompleteness of familiar forms actually arouses our anticipations in almost hallucinatory manner precisely as the experiments with eidetic children describe it. It is again the effort after meaning which leaps ahead of the actually given and completes the shape as we tend to complete a sentence or musical phrase. Hence, perhaps, the increase in the impression of speed and movement which is felt by many observers who compare a frontal photograph of the Discobolos (Pl. 31e) with the less complete and less legible side view (Pl. 31f).

If these devices hint at the possibilities of narrowing the temporal span of the moment represented while enhancing the effect of movement, the victory of the camera in all these methods was bound to make the artist seek for further fields of experiments. The Futurists, of course, with all their glorification of speed and movement, followed the camera rather tamely in their imitations of double exposures. Even Duchamp's famous *Nude descending a Staircase* remains a rather cerebral affair. It needs a great artist to articulate this elusive impression of movement in images of a fresh significance, and here as elsewhere it was Picasso who came up with the most interesting and most varied solutions. In Cubism he played with the idea of various aspects of an identical object, but this programme—so far as it ever was a programme—is little more than a pretext to lead the effort after meaning a hectic chase through a labyrinth of ambiguous facets which both obscure and reveal the still life on the table, recalling the process of vision itself rather than the thing seen.

But it is in some of his later paintings that Picasso seems to have been most successful in giving us a feeling of successive images without sacrificing the meaning of their common core.[32] The *Sleeper turning* (Pl. 32a) is such an example, but his greatest triumph here is, perhaps, the *Girl reading* from the early 1950's (Pl. 32b). The strange ambiguity of beauty and plainness, of serenity and clumsiness in these conflicting aspects has nothing directly to suggest a succession of viewpoints in time, but precisely because they are here held in provisional simultaneity they present a novel and convincing victory over that man-made spectre, the *punctum temporis*.

For whatever the validity of these individual devices and however subjective the effect of movement one or the other may produce in some observers, one thing is certain. If perception both of the visible world and of images were not a process in time, and a rather slow and complex process at that,

[32] C. Gottlieb, 'Movement in Painting', *Journal of Aesthetics and Art Criticism*, xvii, 1958.

static images could not arouse in us the memories and anticipations of move-ment. Ultimately this reaction must be rooted in the difficulties we experience in holding all the elements in our mind while we scan the visual field. Hence even abstract art can elude the static impression at least in those extreme cases which exploit fatigue and after-images to produce a sensation of flicker and make the striations and patterns dance before our helpless eyes. An explana-tion of these phenomena experienced before in the black and white paintings by Bridget Riley[33] (Pl. 32c) is still being sought,[34] but even the first attempts throw a fascinating light on the complexity of visual processes. Experiments such as these are wholesome reminders of the inadequacy of those *a priori* distinctions in aesthetics which were the subject of this paper.

[33] Cf. the catalogue of *Gallery One* for September 1963, with four illustrations and notes by D. Sylvester and A. Ehrenzweig.

[34] Donald M. MacKay, 'Moving Visual Images Produced by Regular Stationary Patterns', *Nature*, 1957, 180, 849–50, and 1958, 181, 362–63; the same 'Interactive Processes in Visual Perception' in *Sensory Communication*, ed. Walter A. Rosenblith, 1961; (with more bibliography).

Note : Since this paper went to press the *Bulletin of the British Psychological Society* has published abstracts of two relevant experimental investigations: A. Crawford, 'Measurement of the Duration of a Moment in Visual Perception', XVII/54, 1964, and B. Babington Smith, 'On the Duration of the Moment of Perception', XVII/55, 1964.

Note 1, p. 201, should be supplemented by Etienne Souriau, 'Time in the plastic arts', *Journal of Aesthetics and Art Criticism*, vii, 1949, pp. 294–307.

WALTER J. ONG, S. J.

Evolution, Myth, and Poetic Vision

. . . They say,
The solid earth whereon we tread

In tracts of fluent heat began,
    And grew to seeming-random forms,
    The seeming prey of cyclic storms,
Till at the last arose the man.

. . . Arise and fly
    The reeling Faun, the sensual feast;
    Move upward, working out the beast
And let the ape and tiger die.—(Tennyson, *In Memoriam,* CXVIII)

THE INFLUENCE OF DARWIN upon the poetic and artistic imagination has become a commonplace, documented by a large assortment of studies from Lionel Stevenson's *Darwin among the Poets* (1932) through Georg Roppen's *Evolution and Poetic Belief* (1956). And yet, surveying the work of the creative human imagination today, one is struck by the slightness of creative drive connected with an awareness of evolution, cosmic or organic. It is not that the poets refuse to accept evolution. They render lip service to it. But it does not haunt their poetic imaginations.

One of the great evolutionary philosophers of our day, Father Pierre Teilhard de Chardin, has been accused of writing often as a poet. But we are hard put to find poets who make creative use of evolutionary insights comparable to Teilhard's. Teilhard faces forward, into the future, as, in its brighter moments, does the rest of our world, permeated as it is with evolutionary thinking. But the poets and artists tend to exalt the present moment when they are not facing the past. There is here certainly some kind of crisis concerning the relationship of the poet or artist to time.[1]

The situation is complicated by the fact that today's poets and artists generally are acutely aware of the continuing development of art itself. The existence of a self-conscious *avant-garde* makes this plain enough. Poetry, together with art generally, has a sense of its own domestic time. But cosmic time, as this has been known since the discovery of evolution, is another matter. Most poets and artists are not much interested in it, even when they are most intently concerned with man, who exists in this time. Writers who do deal with larger patterns of development in time tend to slip into thinly veiled sensationalism, as does George Bernard Shaw in *Back to Methuselah,* or sensationalism not so thinly veiled, as in George Orwell's *1984,* or they handle cosmic time not very successfully, as does Hart Crane, or half-heartedly, as does T. S. Eliot. One feels that, in the last analysis, the poet and artist are not very much at home in an evolutionary cosmos.

The basic issue between poetry and evolutionism is seemingly the need in poetry, as in all art, for repetition. The drives toward repetition show in poetry in countless ways—in rhythm, in rhyme, in other sound patterns, in thematic management and plotting (Joyce plots *Ulysses,* which for all practical purposes is a poem in the full sense of this term, to match Homer, as Virgil in a different way plotted the *Aeneid*). Even the key to all plotting, recognition, is a kind of repetition, a return to something already known.
In "Burnt Norton" T. S. Eliot writes:

> And the end and the beginning were always there
> Before the beginning and after the end.[2]

Finnegans Wake is a serpent with its tail in its own mouth, the *ouroboros*: the last words of the book run back into its first words.

The preoccupation of poetry and of art in general with repetition is shown at its deepest level in the constant resort to the natural cycle of the year: spring, summer, autumn, winter. Mircea Eliade has shown the tremendous drive of this cycle within the human consciousness in his book *The Myth of the Eternal Return*. Indeed, the cosmic myth of the seasons, with its lesser parts, its contractions, expansions, and other variations and projections (the succession of day and night, the imaginary Hindu *kalpa* of 4,320,000,000 solar years, Yeats's elaborate hocus-pocus in *A Vision*), dominates the subconscious so thoroughly that one can speak of it simply as natural symbolism—all nature symbolism comes to focus here—or even as *the* myth, for, in effect, there is no other. Professor Cleanth Brooks, distinguishing interest in history from interest in nature, notes that in modern poets "the celebration of nature is not tied to a cyclic theory."[3] It is my conviction that it need not be. But even in the writers Professor Brooks cites, such as Dylan Thomas, there is a discernible hankering for cyclicism; and in others he cites, such as Wallace Stevens, who shows keen interest in non-cyclic change, one finds less than a wholehearted welcome of a truly historic view. As we shall see, in place of the continuities of history one finds in Stevens rather a discontinuous series of states of chaos, each separately resolved by the imagination, each resolution, in a sense, being a kind of repetition of foregoing resolutions, with no recognizable progress.

In a perceptive study Professor Northrop Frye has recognized this fact, proffering a classification of the archetypes of literature based on the natural cycle of the year because "the crucial importance of this myth has been forced on literary critics by Jung and Frazer."[4] Professor Frye's first phase is the "dawn, spring and birth phase," concerned with the hero, revival and resurrection, creation, and the defeat of the powers of darkness, and having as subordinate characters the father and the mother. This, he states, is the "archetype of romance and of most dithyrambic poetry." The second phase is that of "zenith, summer, marriage or triumph." Here we are concerned with apotheosis, the sacred marriage, and entering into Paradise, and with the subordinate characters of the companion and bride. This is the archetype of comedy, pastoral, and idyll. The "sunset, autumn and death" phase is the third, concerned with the dying god, violent death and sacrifice, and the hero in isolation. The traitor and the siren are subordinate characters, and this phase is the archetype of tragedy and elegy. The fourth and last phase is the "darkness, winter and dissolution phase," with its floods, return of chaos, defeat of the hero—the *Götterdämmerung,* accompanied by the ogre and the witch as subordinate characters. This is the archetype of satire, as instanced in the conclusion of *The Dunciad.*

Waiving questions as to the applicability of the details of this structure to the actuality of poetry and art, we can see that Professor Frye is presenting us here with something on the whole both real and powerful. Moreover, as he himself observes in the same place, the natural cycle not only touches poetry in terms of its themes, imagery, and characters, but also in more pervasive terms, such as that of rhythm itself, which appears essential for art, verbal or visual: "Rhythm, or recurrent movement, is deeply founded on the natural cycle, and everything in nature that we think of as having some analogy with works of art, like the flower or the bird's song, grows out of a profound synchronization between an organism and the rhythms of its environment, especially that of the solar year."

Everyone can recognize the actuality of these rhythms, too. Spring does come back each year. Day succeeds night, and night day. Men are born and die. There are, however, certain problems here in establishing rhythmic patterns. The likening of man's life to a cycle, for example, is based on an all too obvious distortion: there is *some* likeness between the helplessness of an old man and that of an infant, but to mistake one for the other one would have to be out of one's mind —here the cyclic myth has asserted its compelling power in consciousness and made plausible in our assessment of human life a pattern which is really not there: the life of an individual actually ends quite differently from the way it began. One can think otherwise only by blotting out certain facts.

The same is true with regard to groups of men taken as groups. In an article a few years ago, I pointed out in some detail that the likening of the "life" of a nation or empire or of a culture or people or tribe to the life even of an individual man, and *a fortiori* to a perfect cycle in which the end is the same as the beginning, is quite indefensible and utterly contrary to fact, although by leaving out of consideration certain obvious facts, by proper selectivity, a certain analogy, very loose, between a nation and an individual and a much feebler analogy between the history of a social group and circular movement can be made out.[5] But, on the whole, these analogies probably deceive more than they inform. The

Roman Empire "fell" (returned to the starting point from which it had presumably "risen") only in a very loose sense. It "died" only in a very loose sense, too, for it had never really been conceived and born as a human being is. The institutions of the Roman Empire are still all around us and in us, more widespread today than ever before; the descendants of its citizens are extraordinarily active over a greater expanse of the world than ever before, as mankind becomes more and more unified. Much as a circular area, say, a foot in circumference, can be discerned on an absolutely blank blackboard simply by disregarding the rest of the blackboard, so rise-and-fall or birth-and-death patterns can, of course, be discerned in events in the stream of time by proper selectivity. But what do such patterns explain? We like the rise-and-fall pattern probably less because it informs us about what is actually going on in the world than because it is, after all, a pattern, and the simplest pattern of all, imposed on the field of history, noteworthy for its lack of pattern. The attraction of periodicity operates largely from within the human psyche.

What sort of actuality do the cycles of nature have when we view them in terms of what we know of the universe since the discovery of cosmic and organic evolution? In the last analysis, they do not have much. Rhythms are approximations. Perfect cycles, exact repetitions, recurrences of identical starting points, are not really to be found. Although each winter is succeeded by spring, every year is actually different from every other if we look to details. What lengthier rhythms there may be—several years of drought and several of floods —are not really exact cycles, but approximations of cycles which gradually alter. On the whole, the global climate is changing in some kind of linear-style pattern, for the evolution of the earth is progressing toward an end-point quite different from its beginning. In the cosmos as we now know it, there is no real repetition anywhere, for all is in active evolution. One sees repetition only in the rough, where one does not examine more closely. But the universe is being examined more closely all the time. Weather patterns, to stay with our example, are being fed into computers to give us the remarkably accurate forecasting which has developed over the past decade or so. Climatic changes are being studied as they really occur over telling expanses of time, not as impressionistic constructions fabricated out of the limited experiences in one man's lifetime, inaccurately recalled.

Of course, there is a human dimension to the universe, and in the dimensions of one life, rhythms of repetition humanly identifiable and humanly satisfying are to be found. But the human dimension today also includes a great deal of abstract, scientific knowledge—for science is nothing if not a human creation, since it exists only within the human mind. Our abstract, scientific knowledge, which is now entering so thoroughly into planning as to be eminently real as well as abstract, includes a knowledge of the evolution of the cosmos and of life. This means that, in conjunction with an immediate experience of approximate recurrence, we experience also, if we are alert to the world in a twentieth-century way, an awareness of the fact that recurrence does not stand up *in detail*. Quite literally, in the modern physical universe, nothing ever repeats itself. Least of all does history.

216

The classic model for cyclic repetition, when it was rationalized, had been the supposedly immutable path of the sun around the earth. Now we know not merely that the earth moves around the sun, but also that it moves in a path, not circular but elliptical, which is gradually changing its form, in ways which are measurable. The stars are not changeless, but in full evolutionary career. So is our solar system. And the elements themselves are dismembered and reconstituted in the process of cosmic evolution.

One can still project a cyclic model of perpetual repetition upon actuality, pretending that everything now happening happened before an infinite number of times and will happen again an infinite number of times. But study fails to reveal any warrant within actuality itself even for the model. Even if we are living in a so-called "throbbing universe," which expands to a maximum and then over billions of years reverses and contracts to a single, unimaginably hot super-atom only to explode and expand again, all the evidence we have around us from the universe itself suggests that the pattern of events in the second explosion will be different from that in the first. To cap all this anti-repetitiveness is the appearance of human life itself in the cosmic process. For each man is a unique individual, utterly different from his fellows, all of them, no matter how many they are. The difference is not merely genetic. It is conscious, as can be seen in identical twins, who have the same genetic structures but quite different consciousness, the one "I" utterly distinct from the other. Each of us knows he is unique—that no one else experiences this taste of himself which he knows directly, a taste, as Gerard Manley Hopkins put it, "more distinctive than . . . ale or alum." No one in possession of his wits is concerned that one of the other three billion or so persons in the universe today is identical with himself. For each man knows his own induplicability and interior inaccessibility. In simply knowing himself, each knows that his interior landscape is unique and open only to his own mind. With man at the term of the cosmic and organic evolutionary process, we thus are aware of the universe in its entirety as building up to maximum unrepeatability, self-conscious uniqueness, singularity folded back on itself.

With this kind of awareness, what remains of recurrence as a foundation for poetry and art? We are, of course, as we have seen, still acutely conscious of approximate recurrence to a degree: there is, after all, the evident succession of spring, summer, autumn, winter, repeating year after year. But this basic repetition, and all that goes with it, is no longer at the heart of life in the way in which it used to be. It has been displaced. It is now eccentric. A somewhat sentimental account explains the displacement by urbanization and industrialization: large numbers of men now live far from the wilds of nature or the domesticated life close to nature on the farm. But, more radically, the displacement has come about by the intellectual discovery of the cosmic facts, which are known to persons in rural areas as well as in the cities: we live not in a cyclic, perpetually recurring, but in a linear-type time. I say "linear-type" rather than "linear" because time, being nonspatial, is not entirely like a straight line, either. But it is like a straight line rather than a circle in the sense that events in time end at a different point from that at which they begin. (Whether they are really "strung out" like points on a line is another queston: in fact, they are not.) My life at its end is different from what it was at its beginning. The universe, even now, is different from what it was five billion years ago and gives evidence of continued progressive differentiation from its initial stage and all subsequent stages.

The displacement of the sense of recurrence as the dominant human awareness is, I believe, a major crisis, and probably the major crisis, in the arts today. The displacement does not, of course, affect everyone in society equally. The sensibility of millions of persons, even in highly technologized societies, is doubtless still dominated by a feeling for recurrence which is functionally little different from that of their ancestors two hundred years ago, at least in many areas of life. They do feel the spring, summer, autumn, and winter as a real part of themselves. But even they are undoubtedly affected more radically than they are consciously aware by the psychological structures of society today, particularly by the stress on planning, whether economic or social or industrial or international or interplanetary. Planning means the conscious control of mind over the elements in nature and spells the end of the dominance of quasi-cyclic experience. With planning, matters end up differently from the way in which they began. Moreover, with modern technology, the effect of the seasons—basic to sustaining a sense of recurrence—has been blunted in ways which are sure to be telling, if only subconsciously, for all. A heated and air-conditioned building is pretty much the same in summer and winter, and more and more persons, educated and uneducated, are spending more and more time in such buildings. Transportation, formerly so much affected by the movement of the seasons, is more and more independent of this movement. In technologized societies menus are increasingly the same the year round, or can be. On television one can see skiing in the middle of one's own summer and aquaplaning in mid-winter. The difference between night and day, for practical working purposes, has long since disappeared from major areas of human existence. One has to gloss the text "The night cometh when no man can work" to make it comprehensible to a swing-shift worker in an assembly plant. Even the most unreflective are affected by this detachment of life from the rhythms of nature.

A fortiori the poets and artists are affected. And they know it. In accord with their deeply felt desire for up-to-dateness, which is the desire to speak for man in our time and is itself an anti-cyclic or post-cyclic phenomenon, contemporary poets generally will give at least lip service to the eclipse of recurrence as a central human experience. But how far is poetry affected by this lip service? Poets in English and some other languages continue to use rhyme—although it is significant that they no longer use it so often as they once did. They continue to use lines of more or less matching lengths—although again they do so less than they used to. Occasionally, in fits of desperation, they may resort to bongo drums. But here again, although jazz is indeed relevant to modern living precisely because of its apotheosis of rhythm, resort to jazz is regarded more and more as an escape, if a necessary one. Primitive man banged his drums to attune himself to cosmic harmonies. Modern man resorts to jazz to get away from it all.

The real crisis, however, for modern poets occurs in the images of which they can avail themselves and of course in stylistic and structural devices of repetition where these intersect with or otherwise engage the imagery of a poem. The old reliable cosmological imagery of recurrence appears less and less serviceable. What sort of enthusiasm could be brought today, for example, to the creation of a work such as Edmund Spenser's "Epithalamion," where, if we can believe Professor Kent Hieatt's fantastic calculations,[6] the day and year are represented by the twenty-four stanzas and 365 long lines of the poem, the apparent daily movement of the sun relative to the fixed stars is figured in other line-totals, and at one point the ratio of light to darkness at the time of the summer solstice, when the action of the poem takes place, is properly signaled to the reader? One can, of course, cite Joyce's *Ulysses*—but here the relevance of cosmic imagery is indirect. It is maintained by literary allusion rather than by direct feeling for nature. Joyce builds out of Homer, and countless others, not out of "nature" directly. Of course Spenser builds out of other poets, too, for he is filled with literary allusion. But with him cosmology itself is also more directly operative. Milton, here, is a key figure. *Paradise Lost* was built on a cosmology no longer viable in Milton's day, but clung to deliberately by Milton for poetic reasons. My point is that poets and artists generally today are faced with a crisis similar to Milton's, and even deeper than his was. The polarization of literary dispute around the figures of Milton and Joyce in the mid-twentieth century is perhaps symptomatic: both Milton and Joyce face cosmological problems, and both retreat from them.

Awareness of the modern cosmological crisis in poetry has seldom come to the surface of the contemporary sensibility, and a case for modern "cosmic poetry," with some of the marks of the older recurrence-based patterns, has in fact been made in *Start with the Sun,* by James E. Miller, Jr., Karl Shapiro, and Bernice Slote.[7] The authors of this book also show how, more or less in association with the drift to old cosmic themes, another emphasis is capital in many modern poets: the stress on the esthetic moment, on "creativity," on the instant of "epiphany." This emphasis, which has an obvious Coleridgean as well as Symbolist and Imagist background, deserves attention here, for it throws great light on the poet's relationship to the sense of cosmic time itself. Mircea Eliade has shown that the primitive sense of time, particularly of sacred time, involves a psychological need to recover the beginning of things.[8] Early man—and we can assimilate to early man all mankind generally, more or less, until the psychological effect of typography had entered deep into the subconscious and established a new relationship toward records, the past, and time—early man felt time and change as somehow involving degeneration, a moving away from a perfect "time" at the beginning, a time which was really not a time but an extra-temporal condition, the so-called "time" of mythological existence. The events of mythology—for example, Athena's springing from the head of Zeus, Dionysus' dismemberment by the Maenads—were not the sort of things for which one could supply dates. (As has frequently been noted by scholars, the Biblical accounts of origins involve a different, contrasting sense of time, even when the Biblical accounts are obviously influenced by extra-Biblical mythology.)

Time poses many problems for man, not the least of which is that of irresistibility and irreversibility: man in time is moved ahead willy-nilly and cannot actually recover a moment of the past. He is caught, carried on despite himself, and hence not a little terrified. Resort to mythologies, which associate temporal events with the atemporal, in effect disarms time, affording relief from its threat. This mythological flight from the ravages of time may at a later date be rationalized by various cyclic theories, which have haunted man's philosophizing from antiquity to the present. In the wake of romanticism, however, we find a new refuge from the pressure of time in the cult of the here-and-now esthetic experience, the esthetically achieved moment which gives a sense of expanded existence and of a quasi-eternity. Georges Poulet, in *Studies in Human Time,* Frank Kermode in *Romantic Image,* and others elsewhere have elaborated various ways in which this sense of escape from time is managed, from the French writers leading up to Proust on through various American writers: Emerson, Poe, Emily Dickinson, T. S. Eliot, and others. Post-romantic estheticism depends in great part on the sense of this esthetic moment, different from and more valuable than experiences in ordinary time. We find this sense particularly acute in the Bloomsbury esthetic growing out of and around G. E. Moore's *Principia Ethica,* which influenced so typical a modern writer as Virginia Woolf. James Joyce's doctrine of "epiphany," of course, belongs in this same setting. And the influence on the New Criticism is evident: the poem as "object" is assimilated to a world of vision, which is a timeless world by comparison with that of words and sound. An esthetic of "objective correlatives," whatever its great merits, to a degree insulates poetry from time. Up to a point all poetry provides an esthetic refuge from "real" time, but earlier poetic theory, even that expounding poetry as divinely inspired and thus different from ordinary talk, generally lacks this exaltation of a moment of "realization" which is so commonplace today.

The stress on the moment of realization, on epiphany, under one of its aspects, can thus actually be a dodge to avoid the consequences inherent in the knowledge we have that we live in an evolving universe. It can provide a means of escaping from the real—that is, from cosmic on-goingness—a latter-day time-shelter, replacing the primitive's mythological refuge. This is not to say that the older attempts to escape from time have been entirely abandoned. The quest for a lost Eden, the "radical innocence," which Professors R. W. B. Lewis, Ihab Hassan, and others have discerned in American writers particularly, revives some of the old mythological routines. But this quest for a lost Eden, although real enough, must today be looked for closely to be found. Writers do not openly advertise that their creative drives are being powered by a quest for a lost Eden. They often do talk openly about the value of the esthetic moment.

Once we are aware of the psychological issues here, it is possible to discern some fascinating perspectives in modern poetry. Those which we shall here employ are related to Professor Cleanth Brooks's division, already adverted to, between poets preoccupied with history (related to evolutionism) and those preoccupied with nature (related, as we have seen, to cyclicism). But they refine this division further, as I believe. We can view poets in three groupings, not always too neatly distinct, but, given the proper reservations, highly informative concerning the poet's problem of relating to the known universe.

There are, first of all, those poets who are consumed with the imagery of the old cosmic mythology to such an extent that it rather effectively dominates their entire outlook. Such would be, for example, D. H. Lawrence, Dylan Thomas, Lawrence Durrell, and Robert Graves. The suggestion of cyclicism takes various forms here, but common to them all is at least preoccupation with fertility (or its opposite, sterility). Indeed, the present cult of sex (often clearly an obsession) in literature appears from the point of vantage we occupy here to be a flight from time comparable to the fertility ceremonials of primitives, but more desperate because our sense of the evolutionary nature of actuality makes time more insistent today than ever before. Radical innocence is sought more frenetically because we are more aware of its inaccessibility.

In the case of Lawrence, the cult of sex and death—which yields such beauties as "Bavarian Gentians"—is linked with a nostalgia for the past and conscious revivals of old chthonic images, such as the serpent, which were supposed to restore modern man to his lost Eden. Dylan Thomas immerses himself more spectacularly in nature imagery. "Fern Hill" runs on in a riot of time and fertility symbols: apple boughs, the night, time, barley, "all the sun long," grass, sleep, owls, the dew, the cock, "Adam and the maiden," the new-made clouds, "in the sun born over and over," sky blue trades, morning songs, "the moon that is always rising," "time held me green and dying." This stirring poem is a litany of life and death, in its cosmology still of a piece with Lucretius. Lawrence Durrell celebrates the mysteries of sex with a sophisticated neopagan fervor, having little to do with a sense of man's present position in the cosmos he is taking over more and more, although Durrell does have some sense of temporal progression in the evolution of social groupings. Graves protracts what he takes to be ancient continuities into the present.

> Is it of trees you tell . . .
>
> . . . .
> Or of the Zodiac and how slow it turns
> Below the Boreal Crown,
> . . . .?
>
> Water to water, ark again to ark,
> From woman back to woman:
> So each new victim treads unfalteringly
> The never altered circuit of his fate,
> Bringing twelve peers as witness
> Both to his starry rise and starry fall.[9]

Here one notes strong, and doubtless deliberate, suggestions of the old wheel of fortune, so well known to students of the Middle Ages and so revealing of the pagan cyclicism which haunted the medieval mind. Other poets deeply involved in various ways in chthonic, cyclic themes are Edgar Lee Masters and, most of all, Yeats. Indeed, Yeats is so spectacularly and desperately anti-evolutionary that there is little point in discussing him here. But it is worth noting that in *A Vision,* "Byzantium," and elsewhere his cyclicism comes patently and directly from his poetic needs.

The work of poets such as these, deeply involved in sex, fertility rituals, and, by the same token, death, could perhaps be described as Dionysian; and, by contrast, an evolutionary view, which takes full cognizance that history and time do not fold back on themselves but move resolutely forward with the mysterious upthrust evident in the ascent from protozoans to man, could be described as Apollonian. Perhaps all poetry must be in some way Dionysian because of its sources in the subconscious. But one hesitates to make this judgment if only because one suspects that the Nietzschean division into Dionysian and Apollonian is itself the result of a flight from time. Nietzsche's own cyclicism suggests that his thought, whatever its other brilliances, was not relating itself to the full facts of an evolutionary cosmos.

A second group of poets is related to time in another way. These are the poets adverted to above who attempt to solve the problem of time by greater concentration on the pure esthetic moment. In his *Studies in Human Time* Georges Poulet beautifully describes the way in which Emily Dickinson presents in her poetry moments without past and without future except insofar as the future threatens the loss of the moment.[10] Each poem is a moment of experience which releases us from time:

> Safe in their alabaster chambers,
> Untouched by morning and untouched by noon,
> Sleep the meek members of the resurrection,
> Rafter of satin, and roof of stone.[11]

Miss Dickinson does not flee evolutionary time by resort to the seeming endless recurrences associated with a cult of the Earth Mother. She simply dwells in the instant and attempts to protract it. In this, her work is an early example of what would become a regular style, particularly from the Imagists on, a style revived by many poets at the present moment. The cult of the esthetic moment (or epiphany) marks to a greater or lesser degree the poetic performance and beliefs of James Joyce, Edith Sitwell, Conrad Aiken, Wallace Stevens, E. E. Cummings, William Carlos Williams in his more Imagist phases, and countless others. To a greater or lesser degree it permeates the contemporary consciousness from the heights of the New Criticism down to the level of the most unimaginative beatnik writers. Ezra Pound, with his own complicated sense of history, shows its influence, most evidently in his constant cry to "make it new"— although this exhortation has other implications also. In his poetry and poetic theory, Wallace Stevens, despite his predilection for change, bypasses the development of the universe as such and views existence—poetically conveyed— as a series of disconnected esthetic mergings of imagination and chaos. And the newer generation of poets—James Wright, Robert Bly, Donald Hall, Howard Nemerov, John Knoepfle, and others—may repudiate their predecessors on other scores, but they show, if anything, an even more intense devotion to the esthetic moment, often very intimately conceived.

The drift toward the old chthonic fertility cycles (more noteworthy in the Old World poets, at least until very recently) and the retreat into the esthetic moment (discernible on both sides of the Atlantic) are complemented by a third tendency in modern poetry, a disposition actually to accept linear-type change and even to demand it as a condition of poetic activity. This disposition is more marked among American poets than among British and Irish, a fact which is of course related to the nature of the American experience. Whitman is obviously a striking expositor of this experience, with his attitude of total acceptance toward being and his sense of a dynamic present, diverging toward past and future and uniting and equalizing them. Probably more than the somewhat doctrinaire and clinical acceptance of evolutionism which one meets with in early British writers such as Tennyson, George Bernard Shaw, and H. G. Wells, or even Swinburne, Whitman's sense of participation in the ongoing work of the universe appears to acclimate evolutionism to the poetic world. But does it really succeed? Poulet is quite right in noting that Whitman's is "an enunciation, at once successive and cumulative, of all that has been, and of all that will be."[12] "The universe is a procession with measured and perfect motion," Whitman announces.[13] But, unlike Péguy's comparable procession, which as Poulet again explains, has a termination, Whitman's procession simply advances, occupying worlds and times, but never changing anything, never getting anywhere. In fact, in Whitman we find little if any attention to the inner dynamism of evolution itself; what Teilhard has called the "inwardness" of things, the drive within the evolutionary process which moves from the externally organized original cosmos to the cosmos known and more and more controlled from the interior of man's person,[14] is missing from Whitman.

This is not to say that there is no historicism at all in Whitman. Whitman comes off one of the best in his awareness of the one-directional process of history, for his sense of a dynamic present, diverging toward the past and future and uniting and equalizing them, as well as his sense of the uniqueness of the individual imply a sense of the evolutionary, essentially nonrepetitive movement of time.[15] And yet, Whitman, too, is trapped by the old cyclicism, as, for example, in "Song of the Answerer":

> They bring none to his or her terminus or to be content and full,
> Whom they take they take into space to behold the birth of the
>   stars, to learn one of the meanings,
> To launch off with absolute faith, to sweep through the ceaseless rings and never
>   be quiet again.[16]

In the last analysis there is little or nothing in Whitman to differentiate past and future. Whitman's is still a cult of the present moment, temporally expanded, with little real anguish. For him the present does not grow out of the past, nor the future out of the present. Past, present, and future simply coexist—and all too peacefully. The universe and Whitman's appetite, as Poulet notes, exactly equal one another. How can anything happen when so much bland satisfaction reigns? Whitman has little of the dissatisfactions of the reformer or the future-oriented man.

But if he is not especially concerned about improving things, other American poets more typically are—William Carlos Williams, for example, who insists, dramatically in his *In the American Grain* and by explicit assertion in many other places, that it is the business of the present in America to reconstitute its past and to improve its poetic language and hence its poetic realization of actuality.[17] It is interesting that Williams does not think much in terms of degeneration or decadence (which often reveal a cyclic model in the subconscious): the plight of Americans is not that they have defected from their past but rather that they are only now in a position to lay hold of it reflectively and effectively for the first time, since it now is old enough really to be a past to them. Williams dedicates *Paterson* to this enterprise of recovery, which in a way does look ahead. Yet the time which Williams deals with does not unfold, nor does it thrust forward. The present is authenticated by the past and the future lies as a potential in past and present, but there is little adventure in facing what is to come, little sense of unattained horizons ahead. Such a sense, of course, is not necessary for the writing of poetry, but it would seem to be something which could be included in poetic awareness.

One discerns comparable attitudes in Hart Crane. Crane's vision, conceived in *The Bridge,* is born of his sense of his own moment in history, in time, at the dawn of the machine age. His reactions are not querulous, but positive, like those of Whitman, whom he eulogizes. Crane's confident assertion of faith in the future of industrial America hints at a feeling for linear, evolutionary time. But his compulsion to create the "American myth" drives him toward more cyclic views to fulfill his need for a pattern, and we find in the "Ave Maria," for example, a fascination with the old cosmic movements and with cyclic patterns in a variety of forms:

> Of all that amplitude that time explores,
> . . . . :
> This disposition that thy night relates
> From Moon to Saturn in one sapphire wheel:
> The orbic wake of thy once whirling feet,
> Elohim, still I hear thy sounding heel!
>
> White toils of heaven's cordons, mustering
> In holy rings all sails charged to the far
> Hushed gleaming fields and pendant seething wheat
> Of knowledge,—round thy brows unhooded now
> —The kindled Crown! acceded of the poles
> And biassed by full sails, meridians reel
> Thy purpose—still one shore beyond desire! [18]

The fascination with cyclic patterns echoes in the last line of "To Brooklyn Bridge": "And of the curveship lend a myth to God."

Crane's representation of history is more interiorized than Whitman's expansive canvases, but his quest for a stabilizing myth, a symbolic structure which will somehow catch the historical process in poetic toils, draws him back at times into something like primitive cyclism. Crane had read Oswald Spengler. At other times, perhaps under the influence of P. D. Ouspensky, he retreats from the flow of time into a mythical eternal present which alone exists but is parceled out to man piecemeal.

Crane is typically American in his determination to try to make poetic sense of history. Other Americans show a similar concern. Allen Tate, Robert

Penn Warren, William Carlos Williams, Archibald MacLeish, and Robert Lowell, for example, have felt compelled at least from time to time to build poetry around historical events which have appeared to them as part of their own life-worlds—the Civil War for Tate and Warren, the New Jersey city of Paterson for Williams, for MacLeish American *miscellanea,* New England for Robert Lowell. All these poets evince a distinctly open-end or linear-type view of time. They are helped by the fact that the American past they turn to is a recorded, truly historical past, free to all intents and purposes of prehistory and of prehistory's cyclic tow. (The exception which must be made for the native American Indian prehistory is relatively minor.) Another American, Robert Frost, shows the same open-endedness in his own less explicitly historical, more anecdotal concerns. There is little if any mythical reconstruction in Frost. No cyclic nostalgia shows, for example, in the typically courageous, forward-looking poem "An Old Man's Winter Night." Nevertheless, in the particular perspectives we are considering here, it appears that the achievements of these poets are often limited. Their historical mood is predominantly retrospective. It may seem strange to suggest that history can be anything other than retrospective, and yet we know so much history now that we rightly feel the knowledge of the past driving us into the future. I am not saying that these or other poets should be obliged to treat history otherwise than as they have, for they have done exceedingly well in following each his own genius. Nor do I intend to suggest that anyone should opt for a fatuous view of pure progress as man's destiny in his earthly future. I am only saying that these poets cannot be cited as having caught up in their poetry the entirety of present-day man's real time sense.

Even Pound and Eliot, whose personal and poetic journey from the United States back into Europe was a quite conscious re-entry into history, have not provided a point of view in which one can assimilate a full historical and evolutionary vision to a poetic one. Pound piles historical incident on historical incident. His "Cantos" read as a vast pastiche of eyewitness accounts, overheard conversations, and reflections from everywhere out of the past, with Ecbatana and the ancient Near East jostling what Pound in "Canto XXVIII" styles "solid Kansas." But the impression one gets is not of the development of history so much as it is of a present in which all this history is caught up and somehow moved out of time. "Time is the evil. Evil," "Canto XXX" cries. Eliot's great essay on "Tradition and the Individual Talent," with its sensitive description of the relationship of past, present, and future, provides one of the purest examples of truly historical thinking in our century, and the line from "Burnt Norton" which states, "Only through time time is conquered," is a gnomic expression of the condition of both history and transcendence. And yet, the same "Burnt Norton" opens with a quotation from Heraclitus which states, "The way up and the way down are one and the same," focuses, especially in its part II, on the image of whirling movement ("There is only the dance"), and concludes with the lament, "Ridiculous the waste sad time/Stretching before and after." It is noteworthy that the Heraclitian fragment, "The way up and the way down are one and the same," strongly suggests cyclic fatalism (return to point of departure or inability to leave it) and by no means says the same thing

225

as does Eliot's much advertised other source, St. John of the Cross, or the Gospel source on which St. John relies, "He who exalts himself will be humbled, and he who humbles himself will be exalted." The words of Jesus incorporate a dialectical movement missing in this somewhat paralyzing quotation from the Greek sage. All in all, in his poems and plays (for example, *Murder in the Cathedral,* Acts I and II) Eliot interlaces references to historical, evolutionary time with references to cyclic patterns so frequent and intense as virtually to immobilize the historical. Geoffrey Bullough has pointed out Eliot's preference for "formal patterns" over Bergson's open-ended *élan vital.*[19]

In a sense the point thus far made in this study might be seen as predictable. The poetic theorists from Aristotle through Sir Philip Sidney and beyond always knew that poetry and history were at root incompatible—despite the fact that, as we are well aware today, the poetic imagination has often been stimulated by historical events, proximate or remote. But the point here is precisely that such theory is no longer adequate. The incompatibility of poetry and history is today a more desperate matter than it used to be. A sense of history, seen as evolutionary development, has now become an inevitable dimension of all reflective human existence, and if the very feel for evolutionary development is unassimilable by poetry, then poetry cannot compass one of the most profound and intimate of modern experiences.

A sense of history, which is of a piece with a sense of an evolutionary cosmos, is a sense of the present as growing out of a past with which we are in some kind of verifiable contact, and a sense that the present differs from this past with which it connects and that the future will differ from both present and past. It is a sense of continuity and difference, each reflecting the other, such as Eliot so well expounds in "Tradition and the Individual Talent." We have seen the basic reason why such a sense poses a problem for poets: it undercuts structures dear to them, first by downgrading recurrence as such, making what repetition there is only approximate and somewhat incidental, and secondly by making the present not only a present but also a sequel and a prelude. The problem may not appear pressing when we experience only a single poem, but when we look at the entire body of work of a poet, either in its larger themes or images or in the theory which it at times consciously—perhaps often too consciously—shows forth, the problem, as we have seen, is urgent indeed. A significant drift toward either cyclicism or the isolated moment is unmistakable in modern poetry.

The poet has always been ill at ease, to some degree, in the world of actuality. Poetry is imitation, as the ancients well knew. Admittedly, poetry as such cannot be history. But it must be human, and the urgent question today is whether it must write off the modern experience of evolutionary historicity, whether it can even talk about this experience without betraying itself as poetry. This unresolved question, I believe, is what, deep in the subconscious, in great part underlies the *malaise* of poets and their friends today, occasioning the complaint that poets are outsiders more than they used to be, discarded by "modern society," a seemingly unrealistic complaint, since it appears probable

226

that never have poets been more read and more courted than in our present technological United States. The basic question is: Can poetry face into continuous nonrecurrence as such and assimilate it without distorting it? Can it be that the poet (and the artist generally) feels himself an outsider today less because he has been actively expelled from modern society than because he has failed to make his own one of its deepest insights, its sense of historical time and its drive into fulfillment in the future?

We have noted above the American poets' share in the American sense of drive into the future. This sense holds some promise of change. Further promise of change is to be found in the Christian world view itself, which calls for specific attention here because it has been the source of so much of modern man's sense of history. For the Christian, both the universe and the life of the individual man end in quite different states from those in which they began. Time makes a difference. Time tells. Christian teaching urges no one to try to recover a lost Eden. Salvation lies ahead, at the end of time. And Adam's sin, which drove man from the Garden of Eden, is even hailed in the Holy Saturday liturgy of the Roman Catholic Church as *felix culpa,* "happy fault," because it gave God occasion to send His Son Jesus Christ to redeem man. The promise of the future is thus greater than that of the past. Christian (and Hebrew) teaching underlines the nonrepetitiveness of actuality and by the same token the importance of the unique, unrepeatable, human self, the human person. Christianity, like evolutionary thinking, is anticyclic.

Many of the modern poets who espouse an open-end view of time also give evidence of more or less explicit Christian influence—Allen Tate, Robert Lowell, Richard Wilbur, and W. H. Auden would be examples in point, although I do not believe that any of them has fully solved the problem of assimilating our modern sense of time to the artistic medium. There is however another poet generally classified as modern who is especially worth looking into here for the directness—and precocity—with which he has faced into the problem of time, historicity, and the human person living in time. The grounds on which he faces the problem may be too explicitly Christian to solve the problem for some. Yet there is, I believe, something to learn from him. This poet is Gerard Manley Hopkins, an artist who, although he apparently had read little if any Darwin, is still, I believe, more at home in history and in an evolutionary cosmos than most other modern or near-modern poets, although he is not quite aware of his own entire at-homeness here. His Catholic dogmatic background simply fitted him for an evolutionary time-sense despite the fact that the initial steps toward evolutionary thinking caused no little consternation in Catholic and other religious circles.

The key passages in Hopkins for our present purposes are in "The Wreck of the Deutschland." In this poem Hopkins is dealing with the significance of a horrifying event, a wreck in which a German ship, the *Deutschland,* outward bound from Bremen, foundered on shoals in the North Sea during a storm and was stranded for thirty hours without help, with great loss of life and with the most horrible suffering and distress. In one rescue incident, a seaman, lowered on a rope from the rigging to help a woman or child drowning on the deck, was dashed by a wave against the bulwarks and decapitated. The next morning, according to the *Times* report, "when daylight dawned, his headless body, detained by the rope, was swaying to and fro with the waves."[20] Among the details which he picked up from the *Times* accounts, Hopkins focuses on one particularly: "Five German nuns, whose bodies are now in the dead-house here, clasped hands and were drowned together, the chief sister, a gaunt woman 6 ft. high, calling out loudly and often, 'O Christ, come quickly!' till the end came."[21] The central movement in Hopkins' thought in his poem turns on his inquiry into what this nun meant in her cry, "O Christ, come quickly!" He explores many possibilities—was she asking for rescue? For death as a relief for herself and all those around?—and finally settles for the cry as one of recognition and acceptance. This horrible visitation, this agonizing, not even private but involved with the agony of all those around her, was the real advent of Christ Himself in this nun's life: here she would meet Him in her death, and she called out for Him to come and take her "in the storm of his strides." She sees Christ not as an avenger, but as God, her Lover, and in his love as "the Master/*Ipse*, the only one, Christ, King, Head:/He was to cure the extremity where he had cast her;/Do, deal, lord it with living and dead." This was the point—unknown until now—to which her life had been building up, and she was ready, for she had known that God's coming need not be gentle, that He is present not only in "the stars, lovely-asunder" or in "the dappled-with-damson west," but in all the events in history, even the most horrible, out of which He can bring joy. Hers was a faith which could see God in everything—in disaster as well as joy, indeed most of all in her own death—and never waver in its confidence in Him. Had not St. Paul asked in the *Epistle to the Romans* (8:35, 37), "Who shall separate us from the love of Christ? Shall tribulation, or distress, or persecution, or hunger, or nakedness, or danger, or the sword? . . . But in all these things we overcome, because of him who has loved us."

What we note here is a sense of history at perhaps its highest possible pitch. Hopkins, as we find in his theoretical observations, was devoted to the "instress" of things, to uniqueness itself, to what made each thing itself only, other, different from all else. His poetry everywhere testifies to the intensity of his love for variety, for "all things counter, original, spare, strange," as he puts it in "Pied Beauty." Hopkins connected his interest in the uniqueness of things with the thought of his thirteenth-century predecessor at Oxford, Duns Scotus, but interest in the unique was beyond a doubt far more intense and explicit in the post-romantic Englishman than in his medieval compatriot, who was necessarily far less sensitized to history by his age than Hopkins by his. Hopkins, in fact, is clearly a proto-existentialist in his preoccupation with the singular and the singularity of existence, with "my selfbeing, my consciousness and feeling of myself, that taste of myself, of *I* and *me* above all and in all things, which is more distinctive than the taste of ale or alum."[22] His sonnet "As kingfishers catch fire, dragonflies draw flame" announces a kind of self-definition in action: "Whát I dó is me." But his fascination with the unique and his sense of historicity is shown perhaps most strikingly by the way in which in the "Deutschland" he has fixed on the consciously accepted death of a human being—the utterly unique culmination of an utterly unique existence—as the very focus of existence and meaning.

He relates this death to the action of God's grace—the free gift of God which establishes the unique relationship between each unique individual and God. But grace itself, Hopkins insists, is an historical event. It does not come from heaven, direct from God's existence beyond time. Hopkins knows this will shock but presents it as a central Catholic teaching:

> Not out of his bliss
> Springs the stress felt
> Nor first from heaven (and few know this)
> Swings the stroke dealt—
> Stroke and a stress that stars and storms deliver,
> That guilt is hushed by, hearts are flushed by and melt—
> But it rides time like riding a river
> (And here the faithful waver, the faithless fable and miss).[23]

It is clear from the preceding lines of the poem that the "stress" is God's grace, the pressure he exerts on man's life (firm, delicate, mysterious, in Hopkins' image, like the pressure of the streams trickling down from the surrounding hills which hold the head of water in a well up to its level). This grace, "delivered" through the universe in the violence of storms as well as in the interior movements of consciousness which bring the sinner to repentance and hope, does not come directly from God in eternity ("his bliss") but only in history through Jesus Christ, who was and is both God and man, and as man a real material figure identifiable in actual cosmic time. Hopkins goes on about grace:

> It dates from day
> Of his going in Galilee;
> Warm-laid grave of a womb-life grey;
> Manger, maiden's knee;
> The dense and the driven Passion, and frightful sweat;
> Thence the discharge of it, there its swelling to be,
> Though felt before, though in high flood yet.

The grace at work in the world today comes into the present through the historical life of Jesus Christ—His Incarnation, birth, and, most of all, His passion and death. Even the grace given fallen man antecedent to Christ was given in view of Christ's coming into historical time.

"It dates." This is the scandal. Hopkins' uncanny appreciation of the drives in the human psyche which make it want to dissociate itself and what it values from time is evident in the fact that he recognizes the scandal of time, which creates difficulties even for believers. "Here the faithful waver." For it seems indecent that an Almighty God would tie Himself so firmly into the flux of things, focusing His definitive visitation of man at one single brief period, the lifetime of Jesus Christ, and spreading all out from there. Equally uncanny is Hopkins' deep appreciation of the psychological mechanism of the old cosmic mythologies. Far ahead of his time, writing as though he had read Professor Eliade, he states with precocious insight that myths are nothing less than an attempt to escape from time, to make significance dateless. "The faithless fable and miss." They do not see the movement of grace in life as something that "rides time like riding a river." They try to find meaning by escaping from time.

Written in 1875, only sixteen years after the appearance of *The Origin of Species* and without any discernible direct Darwinian influence, "The Wreck of the Deutschland" actually makes use of a theme assimilable to an evolutionary sense of time, an "open-end," developmental structuring of events more explicitly and downrightly than any other poem of comparable size or importance which I know of since Darwin. The presence of grace has proved, in the Christian sense of history, to be a presence curiously of a piece with the presence which man himself feels in the universe since knowledge of cosmic and organic evolution has shaped his deeper attitudes toward his life-world. Hopkins' open-end view of time is focused in the world of the human person and of grace, which lives in persons, rather than in the more material world of cosmic and organic evolution. To this degree his view remains underdeveloped. Hopkins was not greatly taken with Darwin's discoveries, although perhaps he would have been had be lived longer. But his world is open to them; indeed, it would welcome them, with the sense of the uniqueness of things to which these discoveries can give and have given rise.

Hopkins is certainly not the only poet who is influenced by a Christian sense of God's grace operating in real historical time on persons each of whom is unique. Many other poets, most of them far less consciously, are influenced by the same open-ended historicism, as Professor Brooks has pointed out in *The Hidden God*. Such open-ended historicism is part of the Hebreo-Christian heritage, which in fact was perhaps a necessary condition for Darwin's seeing what he saw: it appears unlikely that a sensibility overconditioned by cyclic views would have been gripped, as Darwin was, by evolutionary patterns. But to say that open-ended historicism and the related evolutionary outlook are at home in the Christian world view is not to say that earlier poets, even the most Christian, entirely succeeded in accommodating a truly Christian sense of time to their poetic sensibilities. Professor Brooks has suggested that "with the breakup of the Christian synthesis, nature and history have tended to fall apart."[24] We have to be careful about imputing to past ages a Christian synthesis. If such a synthesis should include a sense of man's real place in the real physical universe of time and space, as apparently it should, there has been not only no valid Christian synthesis in the past but not even a moderately good synthesis. You cannot have a valid Christian synthesis based on a false cosmology or even on a notably defective one. We must face the fact that earlier cosmologies were both defective and, in many crucial points, false. Nature was never until recent times effectively conjoined with history. The problem today is not to restore an old union but to implement a new one. This problem, the present study suggests, is not particularly distressing to the Christian who understands his heritage in the depths at which it can now be understood. But it is a grievously distressing problem for the poet and artist of our time as poet or artist—whether he be Christian or not—and one from which most poets and artists, consciously or subconsciously, retreat. In other words, it is easier for the Christian as such than for the poet or artist as such to subscribe in the depths of his being to an evolutionary universe. It is also easier, *mutatis mutandis*, for the Jew, since the Old Testament sense of time and the New Testament sense of time are of a piece, although the entry of God into time and history is less intense without the New Testament doctrine of the Incarnation.

The plight of the modern poet and artist is truly extreme. The poet or artist is acutely ill at ease in our present life-world. The earlier life-world belonged to the poets largely because it was so largely constructed out of the archetypal images which poetry and art tend to favor. If to a degree the modern world has rejected the poet, the poet also often has rejected the modern world because it demands a reorganization of his sensibility which is utterly terrifying. If the poet speaks for his age, he tends to speak for those who turn away from the characteristic awarenesses of modern man concerned with history and time.

With some exceptions, in his sense of time and history and of the succession of events, the poet thus has tended to be an aborigine, a primitive. Some maintain that the poet or artist must continue always to be such. I do not believe that he can afford to do so. Of course, no one can prescribe how a poet must speak. If, however, the poet is going to speak for modern man, he is going to have to take into account somehow man's total consciousness, even though this entails a reorganization of his own psyche and of the entire tradition of poetry so drastic as to fill us with utter terror. Very possibly the archetypes in the psyche are themselves in process of being reorganized under pressure of present discoveries. How subconsciously archetypal can archetypes be when they are the objects of knowledge as conscious as that which we bring to them today? Let us be honest in facing the future of poetry and art and man. What will poetry be like ten thousand or one hundred thousand years from now? Will man be able still to live with his once fascinating little dreams of recurrence?

SAINT LOUIS UNIVERSITY

NOTES

1. In another context, but using some of the material used here, I have treated this subject in the study "Myth or Evolution? Crisis of the Creative Imagination," *McCormick Quarterly*, XVIII, Special Supplement (Jan., 1965), 37-56. This previous study was read as a paper at a colloquium on "Myth and Modern Man" sponsored by McCormick Theological Seminary in Chicago on October 22, 1964, with other papers by Paul W. Pruyser, Mircea Eliade, and Schubert M. Ogden. The present study is a revised and enlarged version of a lecture given on May 11, 1964, for the Thirty-First Peters Rushton Seminar in Contemporary Prose and Poetry at the University of Virginia. For material in both these studies I wish to acknowledge help from papers and discussion by members of a 1964 St. Louis University graduate seminar on modern poetry and evolutionism: John K. Crane, Sister Mary Ruth Gehres, O.S.U., Elaine K. Halbert, Judith Hoemeke (Mrs. Gerald A.), Leah Jansky (Mrs. Radko K.), Barbara Lawrence, Lannie LeGear, Young Gul Lee, Catherine Manore, John A. Marino, Sister Mary Joan Peters, O.S.F., Barbara Quinn, Mary Slackford, Sister Dorothy Marie Sommer, C.PP.S., Norman J. Stafford, Doris Stolberg, and Alice Zucker.

2. T. S. Eliot, *Collected Poems 1909-1962* (New York, 1963), p. 180.

3. Cleanth Brooks, *The Hidden God: Studies in Hemingway, Faulkner, Yeats, Eliot, and Warren* (New Haven, 1963), p. 130.

4. Northrop Frye, "The Archetypes of Literature," in *Myth and Method,* ed. James E. Miller, Jr. (Lincoln, Neb., 1960), p. 155.

5. Walter J. Ong, "Nationalism and Darwin: A Psychological Problem in Our Concept of Social Development," *Review of Politics,* XXII (1960), 466-481.

6. *Short Time's Endless Monument: The Symbolism of Numbers in Spenser's "Epithalamion"* (New York, 1960). See also Alastair Fowler, "Numerical Composition in *The Faerie Queene,*" *Journal of the Warburg and Courtauld Institutes,* XXV (1962), 199-239, and the same author's *Spenser and the Numbers of Time* (New York, 1964).

7. *Start with the Sun: Studies in Cosmic Poetry* (Lincoln, Neb., 1960).

8. Mircea Eliade, *The Myth of the Eternal Return* (New York, 1954), *passim;* cf. the same author's *The Sacred and the Profane* (New York, 1957) and *Patterns in Comparative Religion* (New York, 1958).

9. Robert Graves, "To Juan at the Winter Solstice" (1946), *Collected Poems 1959* (London, 1959), p. 212.

10. Georges Poulet, *Studies in Human Time,* trans. Elliott Coleman [with an Appendix, "Time and American Writers," written for the translated edition] (Baltimore, 1956), pp. 345-350.

11. *The Poems of Emily Dickinson,* ed. Thomas H. Johnson (Cambridge, Mass., 1955), p. 151 (n. 216).

12. *Studies in Human Time,* p. 344.

13. *The Complete Poetry and Prose,* ed. Malcolm Cowley (New York, 1948), I, p. 120.

14. Pierre Teilhard de Chardin, *The Phenomenon of Man* (New York, 1959) and *The Divine Milieu* (New York, 1960), *passim.*

15. See *Studies in Human Time,* pp. 342-345.

16. Quoted by Bernice Slote in *Start with the Sun,* p. 238.

17. See, for example, his "Author's Note" contributed to *Modern Poetry: American and British,* ed. by Kimon Friar and Malcolm Brinnin (New York, 1951), p. 545.

18. Hart Crane, *Collected Poems,* ed. Waldo Frank (New York, 1946), p. 8.

19. Geoffrey Bullough, *Changing Psychological Beliefs in English Poetry* (Toronto, 1962), pp. 226-227.

20. "The Historical Basis of 'The Wreck of the Deutschland' and 'The Loss of the Eurydice,'" Appendix [giving the text of the *Times* reports], in *Immortal Diamond: Studies in Gerard Manley Hopkins,* ed. by Norman Weyand (New York, 1949), p. 368.

21. *Ibid.,* pp. 367-368.

22. I have pointed out this existentialist strain in Hopkins in a review in *Victorian Studies,* III (1960), 305-308.

23. "The Wreck of the Deutschland," in *Poems of Gerard Manley Hopkins,* ed. W. H. Gardner, 3d ed. (New York, 1948), p. 57.

24. Cleanth Brooks, *The Hidden God,* p. 129.

Some Books on Time

More technical discussions

P C W Davies 1974 *The Physics of Time Asymmetry* (Guildford: Surrey University Press) 214pp

I Prigogine 1980 *From Being to Becoming* (San Francisco: W H Freeman) 272pp

Wider ranging books and anthologies

K G Denbigh 1975 *An Inventive Universe* (London: Hutchinson) 220pp

K G Denbigh 1981 *Three Concepts of Time* (Berlin: Springer) 180pp

J T Fraser 1975 *Of Time, Passion and Knowledge* (New York: Brazillier) 500pp. A widely ranging book going far beyond science into the arts.

J T Fraser *et al* (ed) 1970, 1974, 1978 *The Study of Time* vol I, II, III (Berlin: Springer). These are larger books covering a wide area.

E Freeman and W Sellars 1971 *The Philosophy of Time* (LaSalle, Illinois: Open Court) 240pp. This book consists of 13 reprinted philosophical articles.

R M Gale (ed) 1968 *The Philosophy of Time* (London: Macmillan) 500pp, paperback. Twenty-two reprinted papers with emphasis on philosophy.

A Grünbaum 1973 *Philosophical Problems of Space and Time* 2nd edn (Dordrecht: Reidel)

D Park 1980 *The Image of Eternity* (Amherst: University of Massachusetts Press) 149pp

P A Schilpp 1974 *The Philosophy of Karl Popper* (LaSalle, Illinois: Open Court). Contains relevant discussions.

J J C Smart (ed) *Problems of Space and Time* (London: Macmillan) 420pp, paperback. Thirty reprinted papers with emphasis on philosophy.

G J Whitrow 1980 *The Natural Philosophy of Time* 2nd edn (London: Nelson)

C Wilson (ed) 1980 *The Book of Time* (Westbridge Books)

Recent works with religious overtones

A R Peacocke 1979 *Creation and the World of Science* (Oxford: Oxford University Press)

A R Peacocke (ed) 1981 *The Sciences and Theology in the Twentieth Century* (Stockfield: Oriel Press)

W Yourgrau and A D Breck (ed) 1977 *Cosmology, History and Theology* (New York: Plenum Press) 416pp. Contains 24 articles on a whole spectrum of problems.

Glossary

Abundance (of chemical elements). The fraction of the mass of an element present in a system divided by the total mass present. It can be estimated for the earth, the universe, etc.

Annihilation operator. See Creation operator.

Anti-kinetic behaviour of a system. The behaviour exhibited by a large system approaching equilibrium when all molecular velocities are instantaneously reversed. The entropy of such a system will then decrease with time. For a large system this can be done only in a computer simulation and not experimentally.

Antimatter. Each elementary particle has associated with it an antiparticle which has the same mass but opposite electric charge (and possibly opposite other characteristics). The electron has the positron and the proton the antiproton as antiparticles. The photon and some other neutral particles are their own antiparticle. Antihydrogen is built from an antiproton and a positron. Antimatter is the analogue of ordinary matter, but composed of corresponding antiparticles.

Asymptotically flat space–time. This is space–time in the absence of gravitational effects (flat space–time) at large spatial distances from gravitating matter.

Big-bang models. These are expanding models of the universe, which, on being traced backwards, approach after a finite time a state of infinite matter density with space collapsed to a point.

Black-body radiation. This is electromagnetic radiation in equilibrium with any body of fixed temperature. It is of the same composition as the radiation emitted by a totally absorbing ('black') body at the same temperature. The radiation is said to have a Planck spectrum (named after Max Planck).

Black hole. A region of space with a matter density high enough for light emitted in the region to be bent back gravitationally so that it cannot leave the region. Its properties depend on its mass M, electric charge Q and angular momentum J, and are independent of the nature of the matter which has fallen into it. This is called the theorem that 'a black hole has no hair'. If $Q = J = 0$ one has a Schwarzschild black hole whose radius is $2.95M/M_0$ km where M_0 is the solar mass. If $Q = 0$ and $J \neq 0$ it is called a Kerr black hole and is rotating. A burnt-out star with $M \gtrsim 2.5M_0$ is believed to undergo an implosion under the gravitational forces to form a black hole, but under certain conditions (e.g. sufficient rotation) this gravitational collapse may be halted.

Boltzmann (transport) equation; Boltzmann model of a gas. See Molecular dynamics.

Bulk viscosity. The property of a fluid which determines how its energy of motion is dissipated as heat.

Cauchy hypersurface. This is a surface in space–time usually given by $t = $ constant, at each point of which initial data is known or assumed, and is sufficient to determine the physics throughout the entire space–time.

Causality. The view that any event depends for its occurrence on other events. These 'other events' are called causes, and the event considered in the first place is called the effect. The causality constraint is less severe than that of determinism.

Chandrasekhar limiting mass (~ 1.4 solar masses). A star of smaller mass has adequate electron gas pressure inside it to arrest a gravitational collapse described under 'black hole'. A star in the range if 1.4 to 2.5 solar masses is expected to collapse to a high density state, called a neutron star, but not to a black hole state.

Chthonic. Of the earth (often used symbolically).

Coarse-grained description of a system. In a fine-grained description (i.e. one which takes the maximum quantum mechanical information into account) the entropy of a thermally isolated system is time-independent. If some of this information is unavailable one can replace it by statistical assumptions and then a coarse-grained description is obtained for which the entropy increases or stays constant (if the system is thermally isolated). No universal method of passing from a fine-grained to a coarse-grained description is available.

Compact surface. A surface which is bounded and closed. A finite plane surface and a spherical surface are bounded, whereas an infinite plane is not. Closure means that all boundary points of the surface are to be part of the surface. For example, a spherical surface with one point removed is not closed.

Copenhagen interpretation of quantum theory. The view that quantum theory provides adequate rules for calculating experimental results. As it works with probabilities it is not a deterministic theory and the Copenhagen interpretation maintains that a deterministic theory does not exist. The Copenhagen interpretation was never clearly defined.

Cosmic censorship. The view that cosmological singularities are *always* surrounded by an event horizon. It is still controversial since the big-bang singularity (and white holes generally) may be regarded as observable in principle.

Creation operator. A mathematical object which in acting upon another such object, called the vacuum state, produces a state with one particle. The removal of a particle is accomplished by an annihilation operator.

Deceleration parameter. A dimensionless measure for the rate at which the expansion of the universe is slowing down.

Density matrix. A quantity of statistical mechanics enabling one to calculate the entropy and statistical averages of quantum mechanical quantities.

Detailed balance. The statement that for a system in equilibrium the transition rate between any two microstates of the system is the same in each of the two possible directions of the transition.

Determinism. The view that laws connect physical events in such a way that future and past situations can be inferred from the present data. It is normally assumed that the laws are exact so that exact input information leads to exact prediction and retrodiction. It is a more severe constraint on natural phenomena than causality.

Distance, positive or negative. See Light cone.

Doppler effect. The shift of the number of oscillations per unit time (frequency) due to recession (red shift) or approach (blue shift) of a source of light or sound. Red shifts are also caused by a gravitational field.

Entropy. A thermodynamic quantity which is a measure of 'disorder' (in a restricted sense) of a large system. It satisfies the entropy law namely that the entropy of a thermally isolated system can stay constant or increase, but cannot decrease.

Ergodic system. An equilibrium system which passes through all the states compatible with its energy in the course of time. For quasi-ergodicity the system must pass merely *sufficiently close* to all these states. Systems composed of non-interacting parts are non-ergodic.

Event. An occurrence at a point in space and at an instant in time.

Event horizon. See Horizon.

Fine-grained description of a system. See Coarse-grained description.

Fine-structure constant. A constant which governs the difference in frequency between certain spectral lines of hydrogen. It is $2\pi e^2/hc \sim 1/137$ where e is the electronic charge, h Planck's constant and c the vacuum velocity of light.

Forces, fundamental. See Interactions, fundamental.

Friedmann models. Certain static, expanding, oscillating or collapsing models of the universe which arise from general relativity theory if simple additional assumptions are made.

Geodesic. The curve of shortest or longest interval between two points in a curved space–time.

Gnomic. Of the nature of maxims or aphorisms.

Gravitational collapse. See Black hole.

H-function. The function for which one seeks to prove the *H*-theorem.

H-theorem. The counterpart in statistical mechanics of the entropy law in thermodynamics. As H is the negative of the entropy, one tries to show that it decreases with time in a thermally isolated large system. One has to define it as a function of, for example, the positions and momenta of the particles of the system.

Hard sphere gas. A reasonably tractable model of a gas in which the molecules are treated as hard spheres which interact only on collision.

Heat death. The belief that the universe will eventually attain a state of uniform temperature, and that life will be extinct long before that point is reached.

Hidden variables. Unknown variables which one imagines can be introduced into quantum mechanics so as to turn its probability statements into deterministic ones. Their existence is controversial.

Homogeneous. Having the same properties at all spatial points at each instant of time.

Horizon. A *particle horizon* for an observer at an event E on his world line is a surface in space–time separating the particles into those which can from those which cannot be seen by him at E. The world lines of the former intersect his past light cone at E, the latter do not. An *event horizon* is a surface in space–time, usually closed, such that information (light signals, particles, etc) can be transmitted across it in only one direction. These surfaces are closed for Schwarzschild black holes.

Hubble law. The increase of recession velocities (as inferred from red shifts) with distance of galaxies and other objects.

Hubble time. The time since the big bang, assuming that the universe expanded at the present rate. It is of the order of twenty thousand million years.

Hyperspace. A many-dimensional space in which each coordinate has a physical meaning so that physical events or processes can be represented in it.

Inertial frame of reference. One in which a body under no forces is at rest or in uniform motion. The special theory of relativity requires that physical laws be the same in different inertial frames.

Interactions, fundamental. There are four fundamental forces which are given below.

Their strengths are given in brackets on a scale in which the strong nuclear force is unity and the name of the particles with which the force is asociated is also given in brackets: 1, Strong nuclear (1, mesons); 2, Electromagnetic (10^{-2}, photons); 3, Weak nuclear (10^{-14}, intermediate bosons); 4, Gravitational (10^{-40}, gravitons). Numbers indicate orders of magnitude only.

Interval. See Light cone.

Irreversible processes. See Reversible process.

Isotropic. Having the same properties in all directions.

Kaons. Particles which are some of the known carriers of strong interactions and which can be electrically charged or neutral. Violations of CP symmetry were first inferred from observations of neutral kaon decay.

Light cone at an event E in a space−time diagram. Region in the diagram accessible to light emitted at *E* (future cone), or region from which light can reach *E* (past cone). Points inside the cones satisfy $s^2 = c^2t^2 - (x^2 + y^2 + z^2) \geq 0$ (*s* being the distance, *t* the time lapse and *x, y, z* the changes in the three space co-ordinates) and their distances to *E* are called time-like and they are often regarded as positive intervals or 'distances'. If the inequality is reversed one is dealing with points outside the light cone whose intervals or 'distances' to *E* are called space-like, and they are often regarded as negative. If equality holds only light can connect the point to *E*, the interval is zero, and the surface is called the light cone or null-cone. For simplicity only a small part of the light cone at *E* is drawn and this is called the local light cone.

Locality. The view that in a measurement on two particles it is always possible to arrange for them to be at such a long distance apart that the two measurements do not affect each other. This leads to the principle that what happens in one space−time region is approximately independent of variables subject to the control of an experimenter in a far-away region separated from it by a space-like interval. This is sometimes called Einstein locality or the principle of local causes.

Local theory. One which does not need to invoke influences at a given point *P* which arise from distant events not connected to *P*, i.e. it needs no 'action at a distance'.

Macrostate. See Phase space.

Metric. A geometrical quantity which is the analogue in space−time of the distance between two points in space. Different metrics involve different functions of position and time and the geometry which is defined by a certain class of metrics is called Riemannian space−time. This is general enough to allow for curvature of space−time which represents the gravitational field in the general theory of relativity.

Microscopic reversibility. The equality for forward and reverse directions of the transition probability per unit time between two (micro) states of a system. It does not always hold.

Microstate. See Phase space.

Molecular dynamics. The subject which treats the behaviour of a gas by following the paths and collisions of individual molecules.

Naked singularity. A cosmological singularity not surrounded by an event horizon.

Negative heat capacity. Arises when heat loss leads to temperature rise (occurs in gravitating systems, for instance in stars).

'No hair' theorem. See Black hole.

Nucleosynthesis. The build-up of the elements from a hot soup of (elementary) particles such as photons, electrons, neutrinos, etc, as the system cools. It occurs in stars and in the first few minutes of the big bang.

Null cone. See Light cone.

Onsager relations. Assert a symmetry between the coefficients which couple the flow of irreversible currents to the forces which act. They hold in the linear domain, i.e. not too far from equilibrium. Named after Lars Onsager.

Penrose diagram. A space–time diagram so deformed as to bring points and light cones at infinity into the diagram. The deformation changes lengths and preserves angles, and is called a conformal transformation. These diagrams are also called conformal diagrams. Infinities are dealt with as follows: photon paths on such diagrams originate and end on null hypersurfaces $\mathscr{I}^-$ and $\mathscr{I}^+$ respectively which represent past and future null infinity. Free material particles originate and end at points i^- and i^+ which represent the infinite past and the infinite future. Points i^0 represent spatial infinity. For an explanation of TIP and TIF (used here on p. 163), p. 216 of reference 7 on p. 176 should be consulted.

Phase space. A many-dimensional space in which a microstate of a system is represented by a point and a macrostate by a small cell.

Planck spectrum. See Black-body radiation.

Quasi-ergodicity. See Ergodicity.

Random phase approximation. A device whereby unknown quantities (phases of wavefunctions typically) can be eliminated by a statistical assumption which enables one to average over them.

Red shift. See Doppler effect.

Reduction of the wavefunction. Prior to a measurement a quantum system has the potentiality of being in any one of several states. After the measurement the system is in just one state. This effect is referred to as the reduction of the wavefunction.

Reversible process. Limits of real processes such that they do not involve an increase in entropy. They involve the system passing through a continuum of equilibrium states in either direction. Real processes in an isolated system involve an increase in entropy and they are called irreversible as they normally proceed in only one direction.

Ricci tensor. See Riemann tensor.

Riemannian space–time. See Metric.

Riemann tensor. The Riemann tensor is a geometric quantity which determines the curvature of space–time at a point. It can be broken down into two constituent parts: the Weyl tensor and the Ricci tensor. Essentially the Ricci tensor determines that part of the curvature which is due to the matter distribution at the point, the remainder of the curvature being determined by the Weyl tensor. Thus, in particular, in an empty region of space–time (vacuum) the Ricci tensor vanishes and the Riemann tensor reduces to the Weyl tensor.

Self-gravitation. The property that a system of many gravitating particles has a gravitational potential energy obtained by summing over the interactions between all pairs of particles. The gravitationally collapsed state has lowest *gravitational* potential energy.

Singularity of a function of a variable x. The function approaches infinity as x approaches some critical value x_0.

Singularity in relativity. Regions in space–time where the energy density is infinite. Known physical laws break down in the immediate neighbourhood of singularities. The initial instant of the big bang and the interior of a black hole are examples.

Space-like. See Light cone.

Space—time diagram. A diagrammatic representation of the space—time of a physical system. A two-dimensional picture is obtained by suppressing one or two spatial dimensions and plotting time vertically. Local light cones are shown at selected points and they make it possible on the diagram to sketch the world lines of light signals, of free and accelerated particles, etc.

Spin. An intrinsic property of elementary particles which measures its state of rotation. It can take on only half integral multiples or integral multiples of Planck's constant, so that the word 'rotation' is here used in a generalised sense.

Statistical mechanics. The subject dealing with the physical behaviour of systems which are large enough to require the use of probability and statistics.

Steady-state model of the universe. A model of the universe in which the overall density is constant despite the expansion. There is no big bang; isolated matter is continuously created throughout space to keep the density constant.

Stosszahlansatz. The assumption of molecular chaos in a gas to facilitate theoretical treatments of conduction phenomena, proofs of the *H*-theorem, etc.

Superluminal particles. See Tachyons.

Symmetry breaking. This occurs if a system originally unchanged by a symmetry operation, is disturbed or altered so that it is then changed by this operation.

Tachyons. Particles which travel faster than light, also called superluminal particles. They have not yet been observed and whether their existence would be consistent with accepted physical theories is still a matter for debate among scientists.

TCP theorem. These three symbols stand for operations which may be carried out on certain physical systems to yield another possible physical system, perhaps even identical to the original one. They are called symmetry operations. T reverses the direction of time, C reverses the sign of the electric charge, and P transforms to the mirror image. They are called time inversion, charge conjugation, and parity. The theorem states that all nuclear interactions are unchanged if T,C and P are applied simultaneously. The weak nuclear interactions (e.g. kaon decay) violate T,C and P but respect TCP.

Tidal forces. Forces of compression or tension set up in a body by an external force which acts differently on different points of the body. If the forces are all the same they are eliminated if the body is allowed to 'fall' freely in the direction of the forces. Tidal forces cannot be eliminated in this way.

Time-like. See Light cone.

Uncertainly relation. Relations of the type $\Delta a \Delta b \geq p$ arising in quantum mechanics where Δa, Δb are measures of the accuracies with which two observables A and B (say) are simultaneously measured and p is a positive constant (typically Planck's constant h). One effect is that Δa can be reduced to zero only if nothing is known about B.

Unitarity. In the present cases the term refers to a mathematical property imposed on an operator which determines the evolution of a system in time. It ensures that the probabilities which enter the problems have the usual properties (i.e. sum to unity).

Weyl tensors. See Riemann tensor.

World line. The history of a particle (or an observer, etc) represented on a space—time diagram.

Index of Names

Italics indicate that author appears in a reference list. Part D (papers 13 and 14) is not indexed.

240

Index of Scientific Subjects and Ideas

Bold numbers refer to entries in the Glossary. Part D (papers 13 and 14) is not indexed.

243

C violation 111
CP violation 6, 145
Cauchy hypersurface 158, 169, **234**
Causal anomaly 16, 18
  in general relativity 23
Causal influence, faster than light 13
Causality 149–55, **235**
Cause–effect relation 37
  an illusion 27–8
Chandrasekhar limiting mass 116, **235**
Charge conjugation
  broken by interaction of gravitation with electromagnetism 111
  in chemical reaction 5
  operator 4, 111
Clock paradox 15, 96–7
Clumping, gravitational 171
  entropy compared with that in Friedmann models 174–6
  as increasing Weyl curvature 173
Coarse-graining **235**
  and black holes 22
  in statistical mechanics 53
Collapse, classical 184
  of wavefunction 13–4, 152
Collapsing universe and the inside of black holes 172
Compact surface 185, **235**
Computer experiment for kaon decay 143–4
Conditional independence, law of 10, 181
Conformal diagram 163, **238**
Consciousness of observer 14–5, 152
Copenhagen interpretation of quantum mechanics 14, 153, **235**
Cosmic censorship 158, 167, 172, **235**
Cosmological
  coincidences 109, 126, 129
  constant 71
  nucleosynthesis 67–70, **237**
  principle 23
Culham Laboratory 68
Curvature
  Ricci 173
  space–time, leading to particle creation 159, 167
  Weyl conformal 173

Deceleration parameter 70, **235**
Demon
  Maxwell's 15–6, 41, 77–8
  pressure 15
Density matrix 187, **235**
Detailed balance 31, 44, **235**

Determinism 26, **235**
Deuterium
  importance for cosmological models 70
  nucleosynthesis of 68
Diffusion, not T-invariant but weakly T-invariant 8
Dimensional analysis of masses in the universe 126–30
Distributions
  familiar and unfamiliar, to be treated similarly 40–1
  known, going over into unknown 41
Doppler effect 19, **235**

Einstein metric, subject to Weyl's geometry 110
Electric charge, independent of time 130
Electromagnetism
  step ladder interaction in 99, 105–7
  T-invariance 3, 23, 38, 105–7, 136, 149
  Wheeler–Feynman theory 105–7
Energy crisis 63
Entropy 76, **235**
  per baryon 170
  of black hole 21, 159, 172, 184–9
  fine-grained 19, 22, 76, **235**
  fluctuation of 8, 50, 143
  in Friedmann models, compared with that due to clumping 174–6
  gravitational 21, 22, 25, 170–6
  happiness as an analogy 119
  high or low, in big bang 21, 170–6
  increase and coarse-graining 10, 76
  increase as information is lost 41
  increase in oscillating universe 180–2
  maximum, not possible with gravitation 120–1
  non-extensive 21
  order, emergence of 25–6, 150
  state of maximum 160
  of white hole 159, 187
Entropy law 3, 76, **235**
  does not exclude cyclic world process 125
  reformulated 50
Entropy production in the universe 123–6, 178–82
Ergodicity 32, 184–5, **236**
Escape velocity 19
Event horizon 19, 157–9, **236**
Evolution
  contrasted with heat death 77
  and second law of thermodynamics 26
Extensive variables 115–6

Faster-than-light particles 15–6, 95
Fields and forces, propagation of 97–100
Fine-grained entropy 19, 22, 76, **236**
Fine structure constant 129, **236**
Fluctuation
　black hole, effect of on 160, 162, 184–9
　of entropy 8, 50, 143
　in a gas 74
　giant, and life 79, 178
　setting off oscillating universe 125
　offering an escape from heat death 78
Forces and fields, propagation of 97–100
Fourth dimension 28, 91
Free energy produced by gravitation 178–80
Free will 26
Friedmann models 23, 70, **236**
　contracting, as analogous to gravitational
　　collapse 187
　entropy compared with that due to
　　gravitational clumping 174–6
　with matter and radiation 121–5
　see also Universe and Oscillating universe
Fusion 68
Future and past as primitive concepts,
　responsible for arrow of time 136

General relativity
　causal anomalies in 23
　fourth test of 32
　incompatible with variable G 109
　T-invariance 74, 157
Gibbs' milk-and-ink experiment 53, 57
Gravitating systems
　not in equilibrium 120–1
　maximum entropy not applicable 120–1
　and space–time curvature 109
Gravitation
　and entropy increase 171
　non-saturation of 171
　as responsible for arrow of time 135–6
　as source of free energy 178–80
Gravitational
　clumping 171
　entropy 21–5, 170–6
　red shift 19
　slowing of clocks 19
Gravitational collapse
　as analogous to contrasting Friedmann
　　model 187
　and particle creation, as time inverse
　　processes 164
　and thermodynamics 8. 116

Gravitational 'constant'
　relation to Hubble parameter 126
　variable in time 23, 109–11, 126

H theorem 8, 143–6, **236**
　and suppression of antikinetic behaviour
　　134, 144
Happiness, as analogy for entropy 119
Hard sphere gas 143, **236**
Hawking process 20, 159–64
Heat capacity, negative 21, 171
Heat conduction, not T-invariant but weakly
　T-invariant 8
Heat death 77, 161, **236**
　analogy from Himalayan community 119
　contrasted with evolution 77
　fluctuations as offering escape from it 78
　and theology 79
Helium, nucleosynthesis of 68–70
Hidden variables 13, 153–4, **236**
Hodometer 12, 142
Horizon 15, 157–60, **236**
Hubble's law 65, **236**
Hubble parameter 70–1
　related to gravitational 'constant' 126
Human brain 14, 28, 75–6
Hydrogen, nucleosynthesis of 68–70

Idealisation in physics 37
Inertial frames and special relativity 16, **236**
Information
　gathering and time's arrow 146–7, 151
　loss of, equivalent to gain in entropy 41
　sink, black hole as 20
Initial conditions
　instability of some 9, 57, 143–4
Irreversible process 151, **237**

Kaon decay 6–9, 74, 145, **237**
　computer experiment for 143–4

Lag of matter behind cosmological motion
　178–81
Large number hypothesis 23, 109, 125–8
Laws of physics
　contrasted with boundary conditions 3
　as restricting events 92
Life
　definition of 9, 82
　and giant fluctuations 79, 178
　as not weakly T-invariant 9, 24
Light cone 95–100, **237**
　as restricting propagation of effects 99–100

245